Documentation of the Surface-Water Routing (SWR1) Process for Modeling Surface-Water Flow with the U.S. Geological Survey Modular Groundwater Model (MODFLOW–2005)

By Joseph D. Hughes, Christian D. Langevin, Kevin L. Chartier, and Jeremy T. White

Chapter 40 of
Section A, Groundwater
Book 6, Modeling Techniques

Prepared in cooperation with the

Miami-Dade Water and Sewer Department

Techniques and Methods 6–A40

U.S. Department of the Interior
U.S. Geological Survey

U.S. Department of the Interior
KEN SALAZAR, Secretary

U.S. Geological Survey
Marcia K. McNutt, Director

U.S. Geological Survey, Reston, Virginia: 2012
Version 1.0

For more information on the USGS—the Federal source for science about the Earth, its natural and living resources, natural hazards, and the environment, visit http://www.usgs.gov or call 1–888–ASK–USGS.

For an overview of USGS information products, including maps, imagery, and publications, visit http://www.usgs.gov/pubprod

To order this and other USGS information products, visit http://store.usgs.gov

Suggested citation:
Hughes, J.D., Langevin, C.D., Chartier, K.L., and White, J.T., 2012, Documentation of the Surface-Water Routing (SWR1) Process for modeling surface-water flow with the U.S. Geological Survey Modular Ground-Water Model (MODFLOW-2005): U.S. Geological Survey Techniques and Methods, book 6, chap. A40 (Version 1.0), 113 p.

Preface

A new Surface-Water Routing (SWR1) Process was written for use with the
U.S. Geological Survey (USGS) MODFLOW-2005 groundwater model. The SWR1 Process is
designed to simulate surface-water routing in one- and two-dimensional surface-water features
and surface-water/groundwater interactions. The performance of this computer program has
been tested in models of hypothetical surface-water systems and integrated
surface-water/groundwater systems; however, future applications of the programs may reveal
errors that were not detected in the test simulations. Users are requested to notify the USGS if
errors are found in the documentation report or in the computer program.

Although the computer program has been written and used by the USGS, no warranty, expressed
or implied, is made by the USGS or the United States Government as to the accuracy and functionality of the program and related program material. Nor shall the fact of distribution consti-
tute any such warranty, and no responsibility is assumed by the USGS in connection therewith.

MODFLOW-2005, the SWR1 Process, and other groundwater programs are available online from
the USGS at the following address:

http://water.usgs.gov/software/ground_water.html

Acknowledgments

The development of this work began with many conversations with Virginia Walsh of the Miami-Dade Water and Sewer Department, Kenneth Konyha, Laura Kuebler, Wasantha Lal, Angela Montoya, Jayantha Obeysekera, and Dave Welter of the South Florida Water Management District, and Robert A. Evans of the Army Corps of Engineers, Jacksonville, Florida. The authors are indebted to Ned Banta, Richard Niswonger, Lenny Konikow, and David Prudic, the developers of SFR1 and SFR2 (Prudic and others, 2002; Niswonger and Prudic, 2005), for developing the reach concept adopted in the SWR1 Process and identifying many of the limitations that result from coupling one-dimensional surface-water models (such as SFR1, SFR2, and SWR1) to a cell-centered groundwater model. The authors are also indebted to Jeremy Decker and Eric Swain for providing the SWIFT2D datasets used in several of the simulations to evaluate the capabilities and limitations of the SWR1 Process.

The authors are also grateful to the technical reviewers of this report, including Paul M. Barlow, Eve Kuniansky, Steffen Mehl, Richard Niswonger, Dorothy Payne, Alden M. Provost, and Eric Swain of the U.S. Geological Survey.

Contents

vi

Figures

Tables

Conversion Factors

Multiply	By	To obtain
Length		
meter (m)	3.281	foot (ft)
meter (m)	1.094	yard (yd)
Flow rate		
meter per second (m/s)	3.281	foot per second (ft/s)
cubic meter per second (m³/s)	70.07	acre-foot per day (acre-ft/d)
cubic meter per second (m³/s)	35.31	cubic foot per second (ft³/s)
cubic meter per day (m³/d)	35.31	cubic foot per day (ft³/d)
Rainfall rate		
millimeter per hour (mm/h)	0.03937	inch per hour (in/h)
Hydraulic conductivity		
meter per day (m/d)	3.281	foot per day (ft/d)
Conductance		
foot squared per day (ft²/d)	0.09290	meter squared per day (m²/d)
Specific Storage		
1/meter (1/m)	1.094	1/yard (1/yd)
Manning's n		
meter$^{1/3}$/second (m$^{1/3}$/s)	1.486	foot$^{1/3}$/second (ft$^{1/3}$/s)
Slope		
meter per meter (m/m)	1	foot per foot (ft/ft)

Abbreviations

ASCII	American Standard Code for Information Exchange
BCF	Block-Centered Flow (Package)
BAS	Basic (Package)
CPU	central processing unit
DIS	discretization (Package)
EVT	Evapotranspiration (Package)
GB	gigabytes
GUI	graphical user interface
GWF	Groundwater Flow (Process)
HEC-RAS	Hydrologic Engineering Center River Analysis System
HUF	Hydrogeologic-Unit Flow (Package)
ILU	incomplete lower-upper factorization
LAK	Lake (Package)
LPF	Layer Property Flow (Package)
LU	lower-upper decomposition (or factorization)
MIC	modified incomplete Cholesky
MILU	modified incomplete lower-upper factorization
NAM	Name (file)
OC	Output Control (Option)
PCG	Preconditioned Conjugate Gradient (Package)
RAM	random access memory
RCH	Recharge (Package)
SFWMD	South Florida Water Management District
SFR	Streamflow Routing (Package)
SVAP	stage-volume-area-perimeter (table)
SWR1	Surface-Water Routing (Process)
UZF1	Unsaturated-Zone Flow (Package)
WASD	Water and Sewer Department

Documentation of the Surface-Water Routing (SWR1) Process for Modeling Surface-Water Flow with the U.S. Geological Survey Modular Groundwater Model (MODFLOW–2005)

By Joseph D. Hughes, Christian D. Langevin, Kevin L. Chartier, and Jeremy T. White

Abstract

A flexible Surface-Water Routing (SWR1) Process that solves the continuity equation for one-dimensional and two-dimensional surface-water flow routing has been developed for the U.S. Geological Survey three-dimensional groundwater model, MODFLOW–2005. Simple level- and tilted-pool reservoir routing and a diffusive-wave approximation of the Saint-Venant equations have been implemented. Both methods can be implemented in the same model and the solution method can be simplified to represent constant-stage elements that are functionally equivalent to the standard MODFLOW River or Drain Package boundary conditions.

A generic approach has been used to represent surface-water features (reaches) and allows implementation of a variety of geometric forms. One-dimensional geometric forms include rectangular, trapezoidal, and irregular cross section reaches to simulate one-dimensional surface-water features, such as canals and streams. Two-dimensional geometric forms include reaches defined using specified stage-volume-area-perimeter (SVAP) tables and reaches covering entire finite-difference grid cells to simulate two-dimensional surface-water features, such as wetlands and lakes. Specified SVAP tables can be used to represent reaches that are smaller than the finite-difference grid cell (for example, isolated lakes), or reaches that cannot be represented accurately using the defined top of the model.

Specified lateral flows (which can represent point and distributed flows) and stage-dependent rainfall and evaporation can be applied to each reach. The SWR1 Process can be used with the MODFLOW Unsaturated Zone Flow (UZF1) Package to permit dynamic simulation of runoff from the land surface to specified reaches. Surface-water/groundwater interactions in the SWR1 Process are mathematically defined to be a function of the difference between simulated stages and groundwater levels, and the specific form of the reach conductance equation used in each reach. Conductance can be specified directly or calculated as a function of the simulated wetted perimeter and defined reach bed hydraulic properties, or as a weighted combination of both reach bed hydraulic properties and horizontal hydraulic conductivity. Each reach can be explicitly coupled to a single specific groundwater-model layer or coupled to multiple groundwater-model layers based on the reach geometry and groundwater-model layer elevations in the row and column containing the reach.

Surface-water flow between reservoirs is simulated using control structures. Surface-water flow between reaches, simulated by the diffusive-wave approximation, can also be simulated using control structures. A variety of control structures have been included in the SWR1 Process and include (1) excess-volume structures, (2) uncontrolled-discharge structures, (3) pumps, (4) defined stage-discharge relations, (5) culverts, (6) fixed- or movable-crest weirs, and (7) fixed or operable gated spillways. Multiple control structures can be implemented in individual reaches and are treated as composite flow structures.

Solution of the continuity equation at the reach-group scale (a single reach or a user-defined collection of individual reaches) is achieved using exact Newton methods with direct solution methods or exact and inexact Newton methods with Krylov sub-space methods. Newton methods have been used in the SWR1 Process because of their ability to solve nonlinear problems. Multiple SWR1 time steps can be simulated for each MODFLOW time step, and a simple adaptive time-step algorithm, based on user-specified rainfall, stage, flow, or convergence constraints, has been implemented to better resolve surface-water response. A simple linear- or sigmoid-depth scaling approach also has been implemented to account for increased bed roughness at small surface-water depths and to increase numerical stability. A line-search algorithm also has been included to improve the quality of the Newton-step upgrade vector, if possible.

The SWR1 Process has been benchmarked against one- and two-dimensional numerical solutions from existing one- and two-dimensional numerical codes that solve the dynamic-wave approximation of the Saint-Venant equations. Two-dimensional solutions test the ability of the SWR1 Process to simulate the response of a surface-water system to (1) steady flow conditions for an inclined surface (solution of Manning's equation), and (2) transient inflow and rainfall for an inclined surface.

The one-dimensional solution tests the ability of the SWR1 Process to simulate a looped network with multiple upstream inflows and several control structures. The SWR1 Process also has been compared to a level-pool reservoir solution. A synthetic test problem was developed to evaluate a number of different SWR1 solution options and simulate surface-water/groundwater interaction.

The solution approach used in the SWR1 Process may not be applicable for all surface-water/groundwater problems. The SWR1 Process is best suited for modeling long-term changes (days to years) in surface-water and groundwater flow. Use of the SWR1 Process is not recommended for modeling the transient exchange of water between streams and aquifers when local and convective acceleration and other secondary effects (for example, wind and Coriolis forces) are substantial. Dam break evaluations and two-dimensional evaluations of spatially extensive domains are examples where acceleration terms and secondary effects would be significant, respectively.

Introduction

Groundwater flow in many surficial aquifers can be strongly affected by interaction with surface water. Furthermore, conjunctive use of both surface water and groundwater is common in many areas as water managers attempt to satisfy current water-use demands and prevent adverse effects to other existing users and downstream ecosystems. For example, groundwater withdrawals, often equating to only a few percent of the total volume of groundwater in storage, can have substantial and adverse effects on the quantity and quality of surface water (Alley, 2007). The contributions of groundwater to surface water are considered an important part of the total freshwater budget for many ecosystems and essential to maintain their current composition and function (Reilly and others, 2008). Conversely, there are many areas of the western United States where surface-water leakage represents a significant source of water for the groundwater system (for example, see Faunt, 2009). Increasingly, recognition that surface water and groundwater function as a single resource makes it difficult to evaluate surface-water and groundwater resources separately and credibly assess the cumulative effects of a specific hydrologic stress.

In areas where surface-water flow is predominantly caused by topographic variation, surface-water flow and aquifer interactions can be represented with MODFLOW–2005 using the Streamflow Routing (SFR1) Package (Prudic and others, 2004) and its SFR2 successor (Niswonger and Prudic, 2005). The SFR2 Package solves for one-dimensional unidirectional channel flow for conditions where stream discharge is a function of the channel slope. Surface-water flow may be bidirectional (adverse slope) in (1) flat areas with water-surface and bottom slopes less than or equal to 1.0×10^{-4} m/m (or ft/ft), water depths less than 10 m, velocities less than 1 m/s, roughness coefficients for natural streams (for example, 0.05 s/m$^{1/3}$; Coon, 1998), and where the hydraulic slope is equal to the topographic slope; or (2) highly managed areas with pumps and control structures. In these settings, the SFR Package will not accurately represent surface-water flow conditions and surface-water/groundwater exchange. Previous versions of MODFLOW included the functionality to simulate bidirectional surface-water flow. Examples include DAFLOW (Jobson and Harbaugh, 1999) and MODBRANCH (Swain and Wexler, 1996). DAFLOW and MOD-BRANCH included earlier versions of the groundwater-flow process and neither is currently supported by the U.S. Geological Survey.

To represent conditions where surface-water flow may be bidirectional and (or) highly managed, a new Surface-Water Routing (SWR1) Package was written for MODFLOW–2005 to supplement existing SFR capabilities. Similar to the SFR2 Package, the SWR1 Process routes surface water based on a solution to the continuity equation. Unlike the SFR2 Package, however, the SWR1 Process can account for backwater (tailwater) effects, bidirectional surface-water flow, and management of surface water using control structures. The SWR1 Process is written in the FORTRAN-95 programming language using a modular style for consistency with MODFLOW–2005 (Harbaugh, 2005).

The SWR1 Process described herein is the product of a cooperative effort between the U.S. Geological Survey and Miami-Dade Water and Sewer Department (WASD). Groundwater is the major source of potable water in Miami-Dade County, but because of the extensive surface-water system in the county, the effects of groundwater withdrawals on local surface-water stages and flows and how these effects impact the regional surface-water system are of concern to WASD and the local permitting agency, the South Florida Water Management District (SFWMD). The surface-water system in Miami-Dade County is a low-hydraulic-gradient system ($\leq 1.0 \times 10^{-4}$ m/m or ft/ft) that uses a series of pumps and control structures (gated culverts and spillways), each having its own operating rules, to move surface water through the system. The SFWMD and local water managers operate the surface-water system to (1) provide flood protection during the wet season (May to October), (2) provide water from the regional system for water supply during the dry season (November to April), and (3) minimize saltwater intrusion during the dry season by maintaining elevated canal stages. Because of the low hydraulic gradient and operable surface-water pumps and control structures in Miami-Dade County, backwater effects can be significant. As a result, the SWR1 Process was developed and tested to have a considerable degree of generality and flexibility. The generalized approach used in the SWR1

Process may be applicable, as part of MODFLOW–2005, to a wide range of hydrologic settings within the United States and elsewhere.

Currently (2012), MODFLOW–2005 (Harbaugh, 2005) is the most recent version of the MODFLOW code. MODFLOW was originally developed in the 1980s (McDonald and Harbaugh, 1988) and has been continuously updated (Harbaugh and McDonald, 1996; Harbaugh and others, 2000). Unless otherwise noted herein, the term MODFLOW refers, to the MOD-FLOW–2005 version of the code. Notations used in equations contained in this report are explained after their first use, and also are summarized in appendix 1.

Purpose and Scope

This report describes the new SWR1 Process for simulating surface-water routing with MODFLOW. The process is intended to provide an efficient means to simulate dynamic stage, flow, and surface-water/groundwater interaction. The surface-water routing approximations implemented in SWR1 are based on a Newton solution of the nonlinear continuity equation for surface-water flow using direct or iterative solvers. Details on the theory and implementation of these numerical methods are provided herein.

This report also describes how the SWR1 Process calculates surface-water flow and surface-water/groundwater interaction, and how SWR1 is coupled to MODFLOW. Tips for designing SWR1 datasets for MODFLOW models are also described in this report. To demonstrate the accuracy of the numerical methods implemented in the SWR1 Process, a comparison is presented between SWR1 results and (1) a reservoir-routing (level-pool) problem by Bedient and Huber (1988), (2) two-dimensional steady and transient simulations using the U.S. Geological Survey (USGS) Surface-Water Integrated Flow and Transport in Two Dimensions Model (SWIFT2D; Schaffranek, 2004), and (3) a transient looped one-dimensional river simulation with control structures using the U.S. Army Corps of Engineers Hydrologic Engineering Center River Analysis System (HEC-RAS; U.S. Army Corps of Engineers, 2008). An example simulation of aquifer-reach interaction is also presented to demonstrate the use of available solution, geometry, control structure, surface-water/aquifer system coupling, and boundary-condition options. The test simulations are relatively simple problems that can be used to verify proper code installation and as templates for setting up new problems.

Process Overview and Highlights

The SWR1 Process was developed to accurately simulate stages, surface-water flows, and surface-water/groundwater interactions in areas where surface-water gradients are small and (or) there is significant management of surface water. An example of a hypothetical study area appropriate for application of the SWR1 Process is shown in figure 1. This hypothetical study area consists of a variety of different land use types, including natural, agricultural, industrial, and urban. The hydrology is characterized by a combination of channel features and open surface-water features, such as lakes, ponds, and wetlands. The hypothetical study area shown in figure 1 is typical of many communities throughout the United States and other parts of the world.

The SWR1 Process was designed so that surface-water flow can be simulated at a temporal scale that is finer than a MODFLOW time step to better represent the timescale of surface-water flows. A simplified form of the Saint-Venant equations (Saint-Venant, 1843) has been implemented in the SWR1 Process based on an assumption that the process would be applied to continuous simulations (hourly to daily time steps) in support of water-resource planning activities. As a result, the SWR1 Process was not intended as a replacement for hydrodynamic models that solve the full Saint-Venant equations and are applied to event simulations (sub-hourly time steps).

The process contains two surface-water routing approximations: a diffusive-wave approximation of the Saint-Venant equations, and a simpler reservoir-routing (or level-pool) approximation. These approximations can be used together in the same simulation, or as alternatives to one another. With both approximations, users divide surface-water features into reaches. A reach is a section of a stream or other surface-water feature that is contained entirely within a particular MODFLOW model cell. A single MODFLOW cell can contain more than one reach, as necessary, to represent the surface-water features. This reach definition is consistent with the reach definition used in the SFR2 Package. Reaches can also be grouped together to form a reach group that can span one or more MODFLOW cells. SWR1 then solves for a single stage in each reach group by balancing all of the inflows, outflows, and storage changes within the reach group until convergence is met. In some situations where there are many connected canals or other surface-water features with low hydraulic gradients, combining individual reaches into groups can be a computationally efficient strategy for representing routing within a surface-water system and exchanges with an under-lying aquifer.

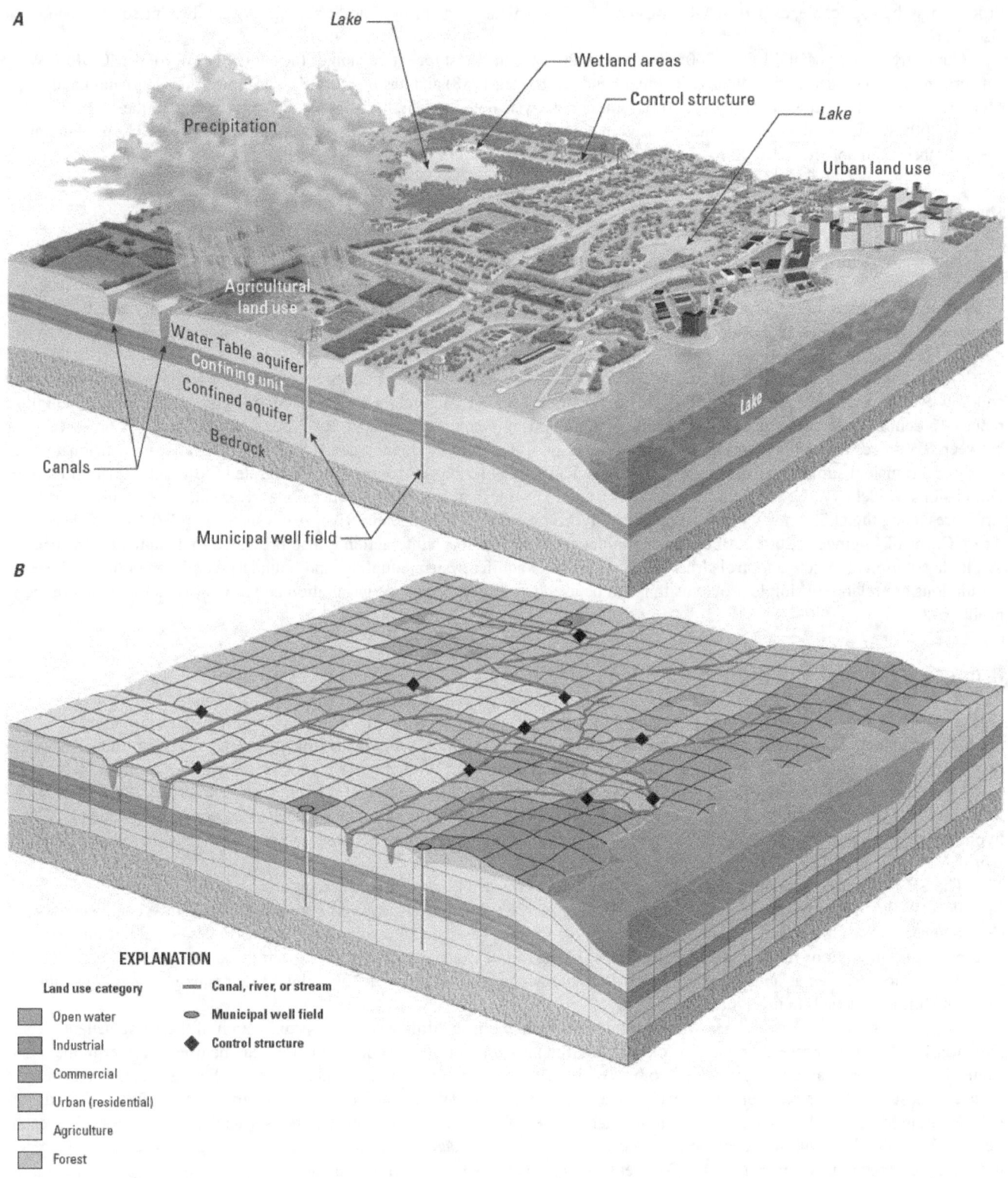

Figure 1. *A*, Conceptual model of a hypothetical study domain showing typical coupled surface-water/groundwater system application that can be addressed using the SWR1 Process; and, *B*, discretized land use categories, surface-water features, aquifers, confining units, and municipal well fields in the hypothetical study domain.

With the diffusive-wave approximation implemented in SWR1, surface-water flows are calculated using a diffusive-wave approximation of the Saint-Venant equations, based on Manning's equation, and the simulated surface-water hydraulic gradient instead of the bottom slope. In most applications of this diffusive-wave approximation, calculations will be done for each reach; reaches will not be combined to form a reach group. Explicitly simulating each diffusive-wave approximation reach means that a dynamic stage will be calculated for each reach. It is possible, however, to improve computational efficiency and avoid some types of numerical problems by grouping several reaches simulated by the diffusive-wave approximation into a single reach group, which can exchange flow with other connected reach groups. With either of the routing approximations, a reach group is the control volume over which the dependent variable (stage) is calculated.

Users assign a geometry type to each reach, which is used by the program to calculate the relation between stage, area, and volume. One-dimensional surface-water features, such as canals and streams, can be characterized by rectangular, trapezoidal, or irregular cross-section geometry. Small isolated lakes and wetlands can be characterized by specified stage-area-volume-perimeter tables. SWR1 also contains a model-based geometry type that assigns an entire model cell to a reach. In this case, the top elevation of the top model layer is used as the reach bottom, and the stage-area-volume relation is calculated using horizontal model-cell dimensions.

SWR1 contains a variety of control structures that can be used to exchange water between reaches. In its simplest form, exchange between reaches may be specified by the user to represent a pump, for example. More complex structures allow the discharge to be calculated as a function of the flow regime and stage difference between reaches. These structures allow submerged and partially submerged weirs and culverts, for example, to be simulated.

A unique and powerful capability of SWR1 is the ability to simulate two-dimensional surface-water flow using model-based reaches. By assigning model-based reaches to a collection of MODFLOW cells, SWR1 will simulate overland flow or surface-water flow within a wetland or lake. This capability was implemented for both the reservoir-routing and diffusive-wave routing approximations. This flexibility allows a multi-cell wetland or lake to be represented efficiently as a single reach group using the reservoir-routing approximation with exchanges with other one- and two-dimensional reach groups controlled using control structures. For situations in which it is important to accurately represent surface-water gradients, the diffusive-wave approximation could be used to calculate a dynamic surface-water stage for each reach and calculate two-dimensional surface-water flow between reaches as a function of stage differences. A combination of one- and two-dimensional surface-water features could be used to represent overbank flooding and realistic channel and overbank roughness differences. Table 1 contains a list of common surface-water features, and the preferred routing approximation for accurately simulating the feature with the SWR1 Process and the associated geometry type.

The SWR1 Process extends the capabilities of MODFLOW to include areas with canals, ditches, wetlands, and lakes, which may or may not be interconnected, as well as to low-gradient rivers and streams. Data input instructions for the SWR1 Process are specified in appendix 2.

Table 1.   Preferred routing approximation and cross section geometry types for common surface-water features simulated using the SWR1 Process.

Surface-water feature	Preferred routing approximation	Cross-section geometry type
River or stream	Diffusive-wave approximation	Rectangular, trapezoidal, or irregular
Canal	Diffusive-wave approximation or reservoir	Rectangular, trapezoidal, or irregular
Multi-grid-cell lake	Reservoir	Model-based reach
Multi-grid-cell wetland	Diffusive-wave approximation or reservoir	Model-based reach
Isolated, sub-grid scale lake or wetland	Reservoir	Specified stage-volume-area-perimeter relation
Spring	Diffusive-wave approximation or reservoir	Depends on characteristics of receiving surface-water feature

Use of SWR1 as an Alternative to Other MODFLOW Packages

The SWR1 Process solves the continuity equation using a general control-volume (finite-volume) approach and can be applied to managed and unmanaged engineered canals and natural settings. As a result, the SWR1 Process can be used as an alternative to some of the existing MODFLOW Packages.

Alternative to River and Drain Packages

SWR1 reaches can be specified to be constant-stage reaches with conductance values that can vary by stress period to control aquifer-reach exchanges. As a result, an SWR1 model dataset can be created to be equivalent to a MODFLOW River (RIV) or Drain (DRN) Package. For example, the DRN Package is sometimes used to represent groundwater springs. The SWR1 Process could be used in the same manner with the added capability to dynamically route spring discharge to connected streams or rivers. Use of the SWR1 Process as an alternative to the MODFLOW RIV Package would give users another way to evaluate aquifer-reach exchange results at scales ranging from individual reaches to groupings of multiple reaches, which may represent reaches between surface-water gages. These capabilities may prove advantageous for gradually transitioning an existing MODFLOW model into a model using dynamic surface-water routing approximations.

Alternative to the Streamflow Routing Package

The SWR1 Process can be used to simulate one-dimensional networks and is analogous to the SFR2 Package (Niswonger and Prudic, 2005) in many respects. It includes similar cross-section geometry options, aquifer-reach exchange options, and lateral boundary options. The SWR1 Process does not have the capability to represent unsaturated flow beneath a stream, which can be simulated with the SFR2 Package. As a result, in cases where the hydraulic-grade line is equivalent to the bottom slope and unsaturated conditions beneath the reach are not significant, the SWR1 and SFR2 Packages would yield comparable results. However, unlike the SFR2 Package (Niswonger and Prudic, 2005; Markstrom and others, 2008), the SWR1 Process offers the capability to represent bidirectional surface-water flow in low-gradient areas. The SWR1 Process also offers the capability to explicitly simulate specific surface-water control structures (for example, culverts, weirs, gates, and pumps).

Alternative to the Lake Package

The SWR1 Process can be used to simulate two-dimensional surface-water features that cover entire grid cells with the reach bottom defined by the top of the MODFLOW model grid. This approach is comparable to the Lake (LAK) Package (Merritt and Konikow, 2000) for lakes connected to the upper layer of a multi-layer MODFLOW model, where the bottom of the upper layer represents the lake bathymetry and the upper layer is inactive outside of the maximum lake extent. However, unlike the LAK Package, the SWR1 Process can be used to simulate two-dimensional surface-water features covering multiple connected grid cells that may have different stages and where exchanges between individual surface-water features (model grid cells) are controlled by bed roughness. An example of such a feature could be a wetland. The LAK Package should be used for deep lakes that cut through one or more model layers; two-dimensional surface-water features in the SWR1 Process are conceptualized as being on top of the model grid.

Dynamic Calculation of Aquifer-Reach Conductance and Coupling to Multiple Model Layers

Similar to the SFR2 Package, the SWR1 Process has been developed to allow dynamic calculation of reach-conductance values used to control aquifer-reach exchanges based on specified reach bed hydraulic parameters and the simulated wetted perimeter of a reach. Dynamic calculation of reach conductance can be important for reach cross sections in which the wetted perimeter increases noticeably with increasing reach stage. The SWR1 Process also includes the option to use the simulated wetted perimeter of a reach and the horizontal hydraulic conductivity of groundwater-model layers or the combination of the horizontal hydraulic conductivity and reach-bed hydraulic parameters to dynamically calculate reach conductance. In some surface-water systems, reach sediments do not impose a significant restriction to flow and aquifer-reach exchanges are controlled by aquifer properties. For example, many of the canals in southern Florida have been excavated directly into limestone units of the Biscayne aquifer and aquifer-reach exchanges are controlled by the hydraulic properties of the aquifer (Chin, 1990; Nemeth and others, 2000; Nemeth and Solo-Gabriele, 2003).

It is also possible to couple a reach to multiple MODFLOW model layers for reaches with defined geometries that intersect more than one groundwater-model layer (fig. 2). Because the reach geometry is defined for all SWR1 reach types, it is possible to vertically distribute the simulated wetted perimeter and hydraulic properties to individual MODFLOW model layers; dry MODFLOW cells are not considered when calculating dynamic aquifer-reach conductance. This option makes it possible to allow the model to dynamically determine which model layers can exchange water with a reach rather than requiring that the

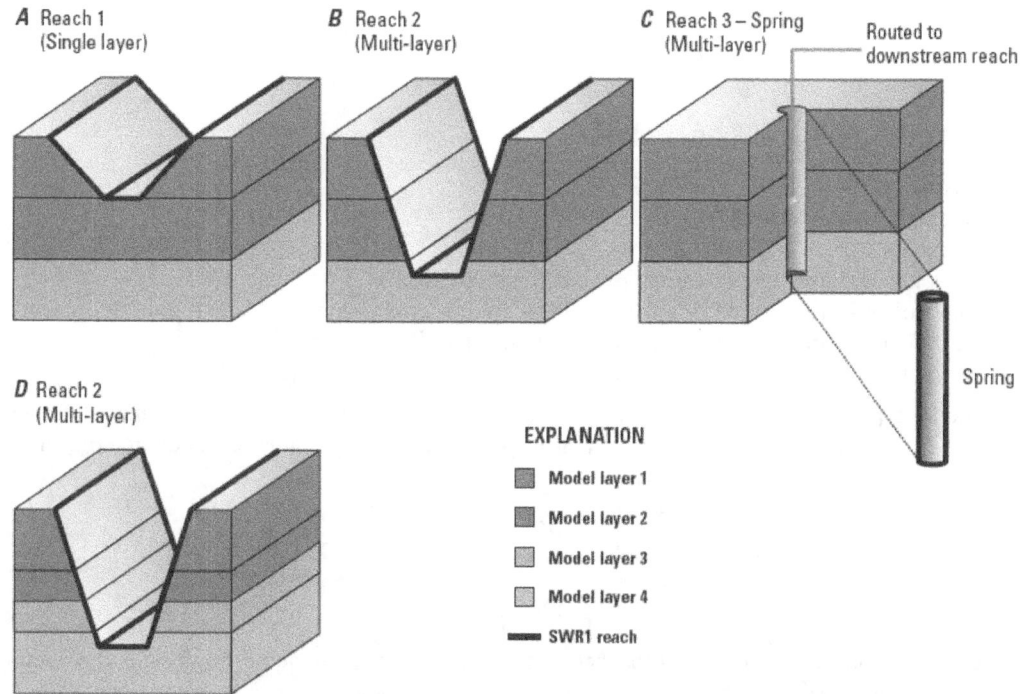

Figure 2.   Example of a SWR1 reach coupled to *A*, a single layer, and *B*, multiple layers. *C*, Example of a SWR1 reach using a specified stage-area-volume table (equation for a cylinder) and representing a spring coupled to multiple layers. *D*, Example of SWR1 reach 2 geometry intersected with a four-layer model instead of a three-layer model.

user specify the model layer with the largest exchange magnitude prior to running a simulation. This capability also can be used to conceptually simulate hydrologic features, such as springs, that are coupled to multiple model layers, and to allow alternative vertical layering schemes to be tested quickly without a need to modify the SWR1 dataset (fig. 2*B–D*).

Dynamic Calculation of Runoff to the SWR1 Process

Similar to the SFR2 Package, the SWR1 Process has been coupled to the Unsaturated-Zone Flow (UZF1) Package (Niswonger and others, 2006). In the UZF1 Package, groundwater discharge to the land surface and infiltration in excess of the product of the saturated vertical hydraulic conductivity and the horizontal model cell area are added instantaneously to specified streams and lakes using the IRUNBND array. SWR1 reaches are coupled to the UZF1 Package by defining IRUNBND values as being equal to the SWR1 reach value plus 100,000. Unlike the SFR2 Package, UZF1 Package runoff is applied to individual SWR1 reaches instead of reach groups (equivalent to SFR2 segments).

Mathematical Representation of Surface-Water Routing

The depth-averaged form of the Navier–Stokes equations, also known as the Saint-Venant equations, are typically used to simulate turbulent flow in one-dimensional channels and on the overland flow plane (for example, Ponce and Simmons, 1977; Akan and Yen, 1981; Lal, 1998). The Saint-Venant equations are developed from the continuity and conservation of momentum equations. The one-dimensional form of the continuity equation for a wide channel, with a constant width, in terms of flow per unit length is

$$\frac{\partial A}{\partial t} + \frac{\partial Q}{\partial x} + q_{PR} + q_{LAT} - q_{EV} + q_{AQ} = 0$$

(1)

where

A	is the cross-sectional area [L^2],	
Q	is streamflow [L^3/T],	
x	is the spatial coordinate [L],	
t	is time [T],	
q_{PR}	is the volumetric flow rate of rainfall per unit length [L^3/TL],	
q_{LAT}	is the volumetric flow rate of inflow and outflow from internal or external lateral sources per unit length [L^3/TL],	
q_{EV}	is the volumetric flow rate of evaporation per unit length [L^3/TL], and	
q_{AQ}	is the volumetric flow rate of aquifer-reach exchanges per unit length [L^3/TL].	

The continuity equation is typically a volume conservation equation in a single-density fluid or is commonly cast as a mass-conservation equation in a variable-density fluid. The one-dimensional form of the momentum equation in terms of flow per time is

$$\frac{\partial Q}{\partial t} + \frac{\partial}{\partial x}\left(\frac{\beta Q^2}{A}\right) + gA\left(S_f + \frac{\partial h}{\partial x}\right) + M_{q_{LAT}} + M_{q_{AQ}} = 0$$

(2)

where

β	is the momentum correction factor that accounts for the effect of a nonuniform velocity distribution in a cross section on the calculation of the overall momentum flux [unitless],	
g	is gravitational acceleration [L/T^2],	
S_f	is the friction slope [unitless],	
h	is the surface-water stage above a datum [L],	
$M_{q_{LAT}}$	is the momentum change due to lateral inflows or outflows [L^3/T^2], and	
$M_{q_{AQ}}$	is the momentum change due to aquifer-reach exchanges [L^3/T^2].	

In this case, stage, h, is the sum of the flow depth (d) and the elevation of the bottom of a reach (z).

Solving the dynamic-wave approximation of the Saint-Venant equations requires solving two sets of simultaneous equations, typically one for stage and another for flow. The Saint-Venant equations can be simplified if the local and convective acceleration terms, the first two terms of equation 2, and momentum changes due to lateral flow and seepage are neglected or, similarly, if defined rating-curve tables or empirical structure equations are used to define flow between calculation points. If these simplifications are valid assumptions, shallow-water flow can be represented using only the continuity equation. Because the SWR1 Process is intended to be used for problems in which acceleration terms and lateral momentum changes are negligible, a simplified form of the Saint-Venant equations has been implemented.

Diffusive-Wave Approximation

The local and convective acceleration terms (the first two terms in equation 2) can be grouped together as

$$M_a = \left[\frac{\partial Q}{\partial t} + \frac{\partial}{\partial x}\left(\frac{\beta Q^2}{A}\right)\right],$$

(3)

where M_a is the momentum change associated with the acceleration terms. If the momentum changes associated with local and convective acceleration terms, M_a, are combined with the momentum change due to lateral inflows or outflows ($M_{q_{LAT}}$) and the momentum change due to aquifer-reach exchanges ($M_{q_{AQ}}$) in equation 2 and divided by gA, then

$$M_0 = \frac{M_a + M_{q_{LAT}} + M_{q_{AQ}}}{gA},$$

(4)

where M_0 is the dimensionless momentum change [unitless]. Combining equation 4 with the one-dimensional momentum equation 2 yields

$$M_0 + \left(S_f + \frac{\partial h}{\partial x} \right) = 0. \tag{5}$$

The relation between the volumetric flow rate between connected reaches (Q_M) and the friction slope (S_f) can be defined using the Manning equation

$$Q_M = \frac{c_M}{n_M} A R^{\frac{2}{3}} S_f^{\frac{1}{2}}, \tag{6}$$

where

c_M is a conversion factor,
n_M is the Manning's roughness coefficient [$T/L^{1/3}$], and
R is the hydraulic radius [L].

c_M is used to convert Manning's roughness coefficients with units of $s/m^{1/3}$ to discharge units consistent with the units of A and R, and has a value of 1.0 and 1.486 for discharge units of m^3/s and ft^3/s, respectively. The hydraulic radius is defined as

$$R = \frac{A}{w_P}, \tag{7}$$

where w_p is the wetted perimeter [L]. Solving equation 6 for the friction slope (S_f) and substituting the result into equation 5 results in

$$M_0 + \left(\frac{n_M^2 Q_M |Q_M|}{c_M^2 A^2 R^{\frac{4}{3}}} + \frac{\partial h}{\partial x} \right) = 0. \tag{8}$$

If the dimensionless momentum change (M_0) is assumed to be negligible, equation 8 can be reduced to

$$Q_M = \mathrm{sign}\left(\frac{\partial h}{\partial x} \right) \frac{c_M}{n_M} A R^{2/3} \left(\left| \frac{\partial h}{\partial x} \right| \right)^{1/2}, \tag{9}$$

where $\mathrm{sign}\left(\dfrac{\partial h}{\partial x} \right)$ is -1 if the calculated gradient is less than zero, and +1 if the calculated gradient is greater than or equal to zero. Rearrangement of equation 9 in terms of a nonlinear diffusion term (D_M) and the friction slope is

$$Q_M = A D_M \frac{\partial h}{\partial x}, \tag{10}$$

where

$$D_M = \frac{c_M}{n_M} \frac{R^{2/3}}{\left(\left| \dfrac{\partial h}{\partial x} \right| \right)^{1/2}}. \tag{11}$$

For two-dimensional overland flow, flow in the x-direction (Q_{Mx}) using equation 10 is formulated as

$$Q_{Mx} = \frac{c_M}{n_M} \frac{d\partial y d^{2/3}}{\left(\left|\frac{\partial h}{\partial s}\right|\right)^{1/2}} \frac{\partial h}{\partial x} = A_x D_{Mx} \frac{\partial h}{\partial x},$$

(12)

where

    ∂y is the model cell width in the x-direction [L] for two-dimensional flow,

    s is the relative spatial coordinate [L] in the direction of maximum local stage slope,

    A_x is the cross-sectional area in the x-direction, and

    D_{Mx} is the nonlinear diffusion term in the x-direction.

The depth, d, can be used because d and the hydraulic radius, R, are equal for two-dimensional overland flow. The equivalent equation for overland flow in the y-direction (Q_{My}) is

$$Q_{My} = \frac{c_M}{n_M} \frac{d\partial x d^{2/3}}{\left(\left|\frac{\partial h}{\partial s}\right|\right)^{1/2}} \frac{\partial h}{\partial y} = A_y D_{My} \frac{\partial h}{\partial y},$$

(13)

where

    ∂x is the model cell width in the y-direction [L] for two-dimensional flow,

    A_y is the cross-sectional area in the y-direction, and

    D_{My} is the nonlinear diffusion term in the y-direction.

Because the momentum equation (eq. 2) has been reduced to a simple equation where flow is a function of stage (h), the equation can be substituted directly into the continuity equation (eq. 1). The resultant modified one-dimensional continuity equation formulated in terms of flow per unit length is

$$\frac{\partial A}{\partial t} + \frac{\partial}{\partial x}\left(AD_M \frac{\partial h}{\partial x}\right) + q_{PR} + q_{LAT} - q_{EV} + q_{AQ} = 0.$$

(14)

Equation 14 formulated in terms of volume per time for an arbitrary control volume with an arbitrary number of connections is

$$\frac{\partial V}{\partial t} + \sum_{i\,1}^{nconn} AD_{M_i} \frac{\partial h}{\partial x} + Q_{PR} + Q_{LAT} - Q_{EV} + Q_{AQ} = 0,$$

(15)

where

    V is the volume of an arbitrary control volume [L³],

    $nconn$ is the number of connections for the control volume (reach group),

    Q_{PR} is the volumetric flow rate of rainfall [L³/T],

    Q_{LAT} is the volumetric flow rate of inflow and outflow from internal or external lateral sources [L³/T],

    Q_{EV} is the volumetric flow rate of evaporation [L³/T], and

    Q_{AQ} is the volumetric flow rate of aquifer-reach exchanges [L³/T].

Equation 15 is the diffusive-wave approximation of the Saint-Venant equations (for example, Akan and Yen, 1981; Hromadka and others, 1987; Feng and Molz, 1997; Lal, 1998), where all dynamic terms (that is, those that can change with time) are defined to be a function of the stage, and can be solved using standard numerical approaches applied to parabolic equations such as the groundwater-flow equation.

Reservoir-Routing Approximation

Lakes, reservoirs, ponds, and detention basins can act as storage features (Chaudhry, 2008) and are commonly referred to, collectively, as reservoirs. The stages in such a reservoir can be considered equalized (a *level pool*) or having a prescribed slope (a *tilted pool*). Equalization is the process of averaging all of the inflows and outflows over the reservoir area in order to calculate an average stage value. Reservoir routing is the process of moving water from one reservoir to the next and is the simplest hydrologic routing approximation that accounts for flow attenuation by storage within a defined storage volume (Zoppou, 1999). With reservoir routing, the Q_M term in equation 15 is further simplified to limit flow to connections between defined reservoirs in the numerical model. In the reservoir-routing approximation, outflow from an arbitrary storage volume is based on a unique, defined relation between stage, volume, and discharge that characterizes a reservoir. Reservoir routing is commonly applied to lakes, reservoirs, ponds, and detention basins, but can also be applied to channels using the assumption that channel routing can be approximated with a series of cascading reservoirs (U.S. Army Corps of Engineers, 1994).

The relation between stage and discharge is defined using tabulated data (that is, a tabulated stage-discharge relation) or empirical control-structure equations (for example, broadcrested weir equation) that define unique discharge relations for defined stage and water-surface gradients across the structure. Changing tailwater (backwater) conditions are not accounted for in cases for which tabulated stage-discharge data are used, but can be represented using empirical control-structure equations; control-structures implemented in the SWR1 Process are discussed in detail in the Surface-Water Control Structure section herein. Tailwater conditions can reduce structure discharge (for example, through submerged orifices and submerged weirs) when downstream stages exceed the control-structure invert elevation, and can be significant during high-intensity flow events. Conceptually, flow between reservoirs can be defined as

$$Q_M = \mathrm{fn}\left(\underbrace{\text{structure type}}, \underbrace{\frac{\partial h}{\partial x}} \right) \text{ or a defined relation} \quad \underbrace{\begin{bmatrix} h_1 & Q_{M_1} \\ h_2 & Q_{M_2} \\ \vdots & \vdots \\ h_n & Q_{M_n} \end{bmatrix}}, \quad (16)$$

(under the fn term: "does account for changing tailwater conditions"; under the matrix: "does not account for changing tailwater conditions")

where structure type is the empirical equation for flow through a specific control structure (for example, broadcrested weir equation or circular-culvert equation).

SWR1 Finite-Difference Approximation

The finite-difference approximation used in the SWR1 Process is developed by adding two additional terms (Q_{BS} and Q_{CS}) to equation 15 to allow for the representation of external boundaries and constant-stage reaches and calculation of complete surface-water budgets for constant-stage reaches, respectively. Q_{BS} is the volumetric flux to external boundaries from surface-water structures that are not connected to another reach. Q_{CS} is the volumetric flux required to balance the inflows and outflows for constant-stage reaches. Constant-stage reaches are commonly used to represent boundary conditions at the upstream and downstream end of reaches not connected to other reaches. A general backward-in-space continuity equation that can be used to simulate a system of reach groups using (1) a diffusive-wave approximation of the Saint-Venant equations, (2) reservoir routing, and (or) (3) constant-stage reaches is

$$\frac{V^t - V^{t+\Delta t}}{\Delta t} + \sum_{i\,1}^{nconn} Q_M^{t+\Delta t} + Q_{PR}^{t+\Delta t} + Q_{LAT}^{t+\Delta t} - Q_{EV}^{t+\Delta t} + Q_{AQ}^{t+\Delta t} + Q_{BS}^{t+\Delta t} + Q_{CS}^{t+\Delta t} = 0, \quad (17)$$

where

V^t	is the volume at the previous time,
$V^{t+\Delta t}$	is the volume at the current time,
Δt	is the current SWR1 time-step length, and
$Q^{t+\Delta t}$	is an individual flux term at the current time.

The finite-difference form of Q_M used to represent the diffusive-wave approximation and surface-water control structures are described in the next section. A conceptual representation of equation 17 is shown in graphic form in figure 3 for a single reach group with a single inflow ($Q_{M(1)}^{t+\Delta t}$) and outflow ($Q_{M(2)}^{t+\Delta t}$) at the current evaluation time step t. Depending on the reach, specific terms may not contribute flow (for example, $Q_{CS}^{t+\Delta t}$) to a reach.

A system of equations for a connected network of surface-water features can be developed using equation 17. Typical numerical methods (finite differences, finite elements, finite volumes) can be applied to solve the system of equations for a specific surface-water network. The numerical solution of surface-water routing applied in the SWR1 Process is described in appendix 3.

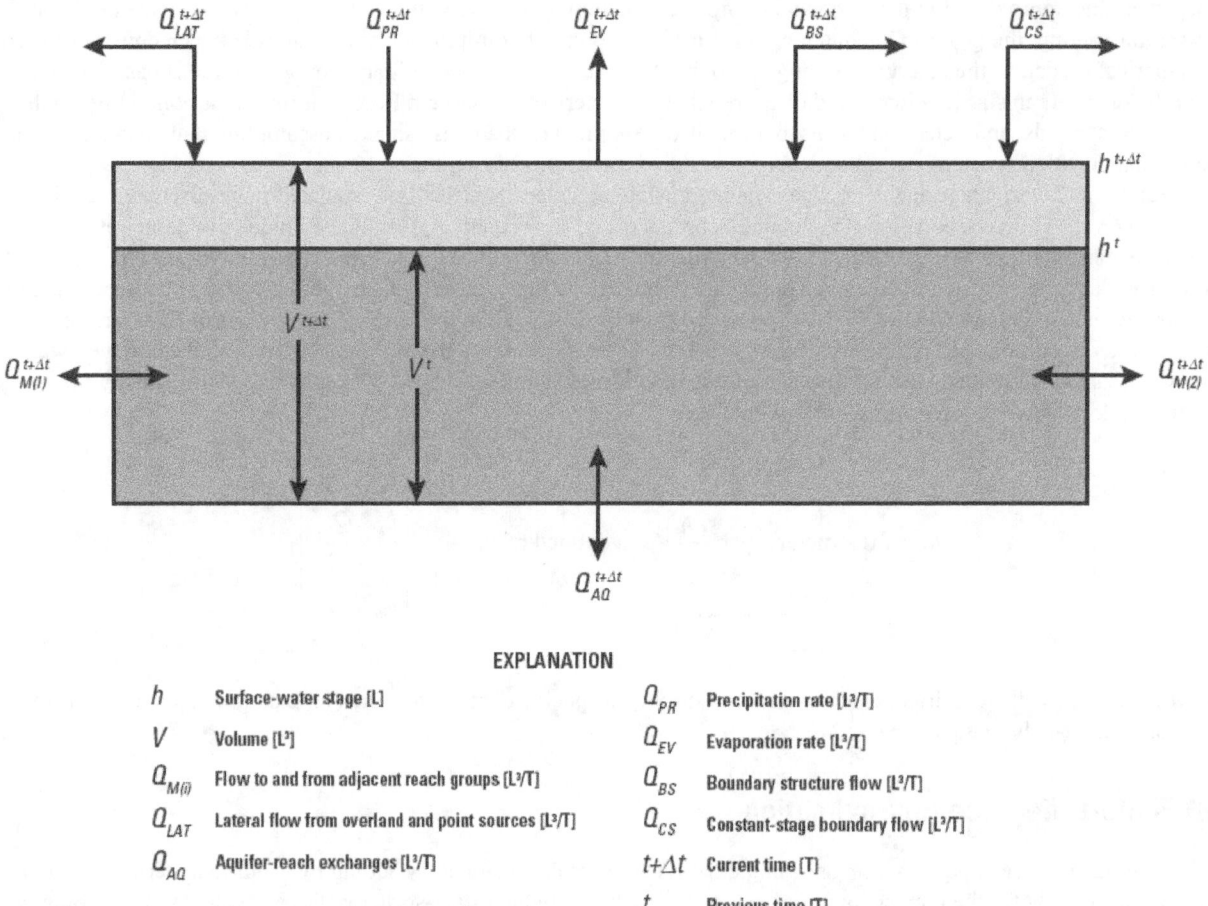

EXPLANATION

h	Surface-water stage [L]	Q_{PR}	Precipitation rate [L³/T]
V	Volume [L³]	Q_{EV}	Evaporation rate [L³/T]
$Q_{M(i)}$	Flow to and from adjacent reach groups [L³/T]	Q_{BS}	Boundary structure flow [L³/T]
Q_{LAT}	Lateral flow from overland and point sources [L³/T]	Q_{CS}	Constant-stage boundary flow [L³/T]
Q_{AQ}	Aquifer-reach exchanges [L³/T]	$t+\Delta t$	Current time [T]
		t	Previous time [T]

Figure 3. Conceptual view of the continuity equation terms defined in equation 17 and solved by the SWR1 Process for a single reach group. The subscripts t and $t+\Delta t$ represent simulated results at the previous and current evaluation time steps, respectively.

Surface-Water Routing (SWR1) Process Concepts and Implementation Details

Surface-water routing within the SWR1 Process is based on a simplified form of the Saint-Venant equations and assumption of piecewise-steady (nonchanging in discrete time periods), uniform (nonchanging in a reach group), and constant-density streamflow, such that during all times, volumetric inflow and outflow rates are equal to changes in storage.

The program is designed to route surface water through a network of one-dimensional channels and two-dimensional surface-water features (which may include rivers, streams, canals, ditches, wetlands, and lakes, and are referred to collectively as reaches in the remainder of the report). The program is also designed to be coupled with MODFLOW and simulate surface-water/groundwater interaction. Streamflow in SWR1 can change direction in reaches, based on the selected numerical option, structure types, and stage gradients. The surface-water model may use time-step lengths that are subdivisions of the ground-water-model time-step lengths. Streamflow can vary from one SWR1 time step to the next in the surface-water portion of the model, but the groundwater head is assumed to be constant for all SWR1 time steps within a single MODFLOW time step.

Spatial and Temporal Discretization

A general approach to surface-water routing that is consistent with the spatial and temporal discretization of MODFLOW has been implemented in SWR1. As previously mentioned, the most basic surface-water routing element in SWR1 is a reach. A reach is a section of a stream, river, canal, or wetland that is associated with a particular finite-difference cell used to model groundwater flow and can have unique (1) routing methods (diffusive-wave approximation routing, reservoir routing), (2) rates of specified inflow and lateral flow (overland flow/runoff), (3) rates of precipitation and evaporation, (4) reach properties (for example, reach bottom elevation, reach bed hydraulic parameters, and stream geometry), and (5) structure types and number of structures. Furthermore, a reach can be specified as active, inactive, or constant-stage. The reach concept used in the SWR1 process is identical to the approach used to discretize surface-water features in SFR1 and SFR2 (Prudic and others, 2004; Niswonger and Prudic, 2005).

A simple example of a surface-water network superimposed on a finite-difference grid of an aquifer that has three columns and three rows is shown in figure 4. This example is based on a similar example described in Prudic and others (2004, p. 3). Each finite-difference cell is designated first by row number followed by column number [for example, cell (R1, C2) refers to row 1, column 2]. In this example, the two approximately parallel streams join and form a stream at the downstream end with a single reach (11) in cell (R3, C3). The first stream has six reaches, with the first reach (1) located in cell (R1, C1); the second reach (2) in cell (R1, C2); the third and fourth reaches (3 and 4) in cell (R2, C2) and separated by a structure; the fifth reach (5) in cell (R3, C2); and the six reach (6) in cell (R3, C3). The second stream has four reaches with the first reach (7) in cell (R1, C2); the second reach (8) is in cell (R1, C3); the third reach (9) is in cell (R2, C3); and the fourth reach (10) is in cell (R3,C3).

More than one reach can be assigned to a model cell, and stream reaches may join together at a junction, or parallel reaches can flow across a cell without joining (fig. 4). A junction is a connection between different reaches. It is not possible to associate a reach with more than one model cell. If multiple reaches are connected to a single model cell, then the groundwater head in this cell is used to calculate aquifer-reach exchanges for all of reaches contained in the cell.

A reach group is user-defined (`IRGNUM`), and can be composed of a single reach or a collection of reaches. Newton methods are applied to reach groups to solve for stages and flows between connected groups. A SWR1 dataset can include reach groups composed of single reaches and collections of reaches. For example, reach group 1 in figure 4 represents a grouping of reaches 1, 2, and 3 into a single group.

Reach groups can be composed of multiple reaches; however, all reaches in a group must use the same routing approximation (either reservoir-routing or diffusive-wave approximation as specified by the `IROUTETYPE` input variable). The total volume and surface area for a reach group is calculated as the sum of the individual reach volumes and surface areas. Flows between connected groups are calculated using either the reservoir-routing or diffusive-wave approximation. Distance-weighted roughness coefficients (`GMANNING`) and cross-sectional areas are used for reach groups composed of multiple reaches using the diffusive-wave approximation to route water between connected reach groups. Calculations of rainfall inflows, evaporation losses, and aquifer-reach exchanges are performed individually for each reach in single- or multi-reach reach groups. The calculated reach group stage is applied to all reaches in a reach group using the equilibrated (level pool) reservoir-routing and diffusive-wave approximation. A user-defined water-surface slope and the calculated reach group stage is used to define reach stages for reach groups using the non-equilibrated (tilted pool) reservoir routing option. Grouping reaches can reduce the computational overhead of the SWR1 Process. Multi-reach reach groups also can be used to combine small reaches, which may result from intersection of the surface-water network and the groundwater-model grid, with larger adjacent reaches to minimize the need for small SWR1 time-step lengths to reduce numerical instabilities.

The SWR1 Process incorporates the same stress-period and time-step concept used in MODFLOW (Harbaugh, 2005). However, because the response times of surface-water systems are generally shorter than those of groundwater systems, the ability to simulate multiple SWR1 time steps within each MODFLOW time step has been implemented. In the SWR1 Process, MODFLOW time steps are generally divided into more than one SWR1 time step.

Similar to other MODFLOW Packages, reach data in SWR1 can vary by MODFLOW stress period. SWR1 reach data that can be varied by stress period include (1) rainfall and evaporation rates; (2) specified lateral flows; (3) reach geometry; (4) reach-bed roughness; (5) aquifer-reach interaction parameters; and (6) the geometry and control criteria of structures.

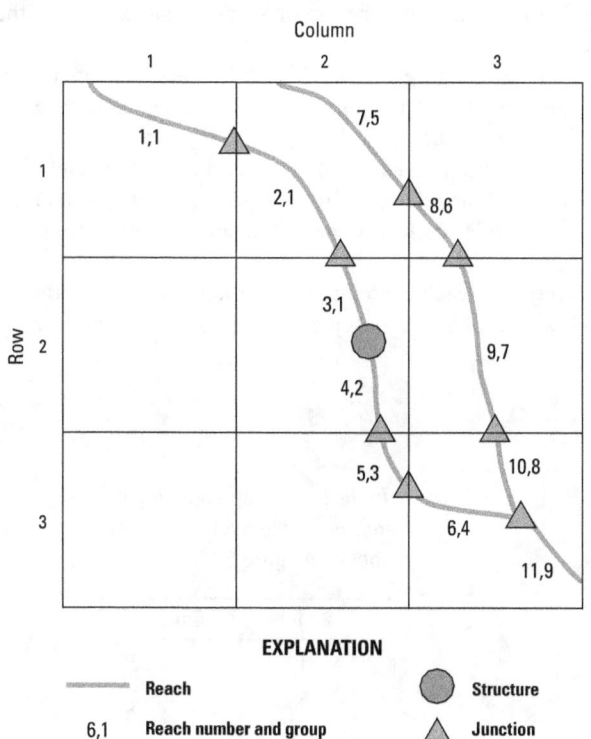

EXPLANATION

⎯⎯⎯ Reach   ● Structure

6,1 Reach number and group   ▲ Junction

Figure 4. Simple surface-water network having 11 reaches, 9 reach groups, 1 structure, and 8 junctions in a finite-difference model grid consisting of three rows and three columns.

Rainfall, evaporation, and lateral flow data can also be specified more frequently than each MODFLOW stress period by using optional external time series files. Surface-water improvements (for example, river dredging or canal widening) can be simulated by modifying reach geometry and conductance coefficients. Seasonal vegetation growth in a surface-water feature can be simulated by modifying the channel bed roughness. Retrofitting control structures can be simulated by modifying the number, geometry, and (or) type of structures in a reach; this modification could include adding a structure where one did not previously exist.

To maintain conservation of volume with time, reach row and column location (IRCH and JRCH) and reach length (RLEN) are defined at the start of the simulation and cannot change thereafter. This approach differs from many of the other MODFLOW stress packages (for example, Well Package and River Package), which allow the number and locations of features to change by stress period. However, it is possible to inactivate or activate reaches during a simulation (ISWRBND). It is also possible to convert reaches from constant-stage reaches to dynamic reaches, or vice-versa, during a simulation. Reaches that are inactivated during a simulation internally maintain the stage and volume specified initially, or from the last dynamic or constant stage time step for the reach to conserve volume. Furthermore, there is no lateral flow (Q_M) to or from inactive reach groups. Storage changes resulting from time-varying constant stages or conversion of dynamic to constant-stage reaches, or vice-versa, are calculated and incorporated into water-budget calculations. With this functionality, it is possible to define the final configuration of a surface-water network and activate selected reaches at specific stress periods to simulate the development of the surface-water system and overcome the requirement that the full configuration of the surface-water network be defined at the start of the simulation.

Reach Connectivity

In the SWR1 Process, reach numbers are used to define the connectivity of reach groups and define reaches connected by control structures. Multiple reaches may be connected as tributaries to single reaches and single reaches can be connected to multiple reaches. Reaches and reach groups do not have to be numbered sequentially from the farthest upstream segment to the last downstream segment. This flexibility in reach and reach group numbering arises from the use of a single matrix to solve the system of equations and the capability to represent bidirectional flow between reach groups.

An example network consisting of five reaches and two junctions is shown in figure 5. In the SWR1 Process, the connectivity for each reach is specified by the user. For example, reach 3 in the simplified network shown in figure 5 is connected to reaches 1, 2, 4, and 5. Table 2 lists the correct connectivity for the five reaches shown in figure 5.

Connectivity at the reach group level follows the reach connectivity. If individual reaches directly correspond to reach groups, reach group connectivity is identical to reach connectivity. Examples of the relation of reaches to reach groups and reach connectivity for (1) single reaches connected to multiple reaches, and (2) multiple reaches connected to single reaches are shown in figure 6.

In the SWR1 Process, structures make a controlled connection between the reach containing the structure and a specified connected reach. Figure 7 illustrates how reaches should be connected if a control structure separates (1) a single reach from multiple reaches, and (2) multiple reaches from a single reach.

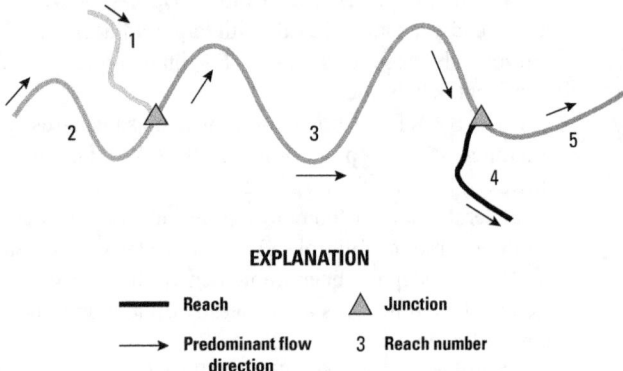

EXPLANATION

⎯⎯ Reach   △ Junction

⟶ Predominant flow   3 Reach number
        direction

Figure 5.   Example of a one-dimensional network with five reaches and two junctions. Different color line segments represent individual reaches.

Table 2.   Reach connectivity for the simple SWR1 network shown in figure 5.

Reach	Connections
1	2, 3
2	1, 3
3	1, 2, 4, 5
4	3, 5
5	3, 4

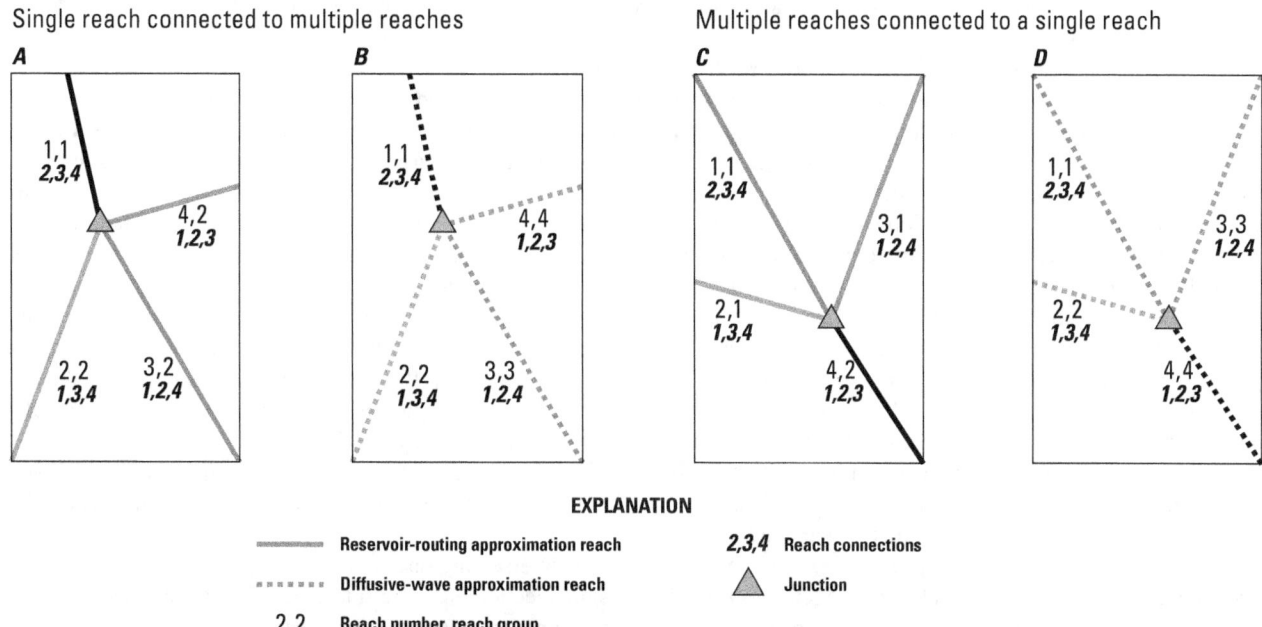

Figure 6. Examples of the connectivity that should be developed for problems where *A*, a single reservoir-routing approximation reach is connected to multiple reservoir-routing approximation reaches, *B*, a single diffusive-wave approximation reach is connected to multiple diffusive-wave approximation reaches, *C*, multiple reservoir-routing approximation reaches are connected to a single reservoir-routing approximation reach, or *D*, multiple diffusive-wave approximation reaches are connected to a single diffusive-wave approximation reach. There is no difference in connectivity between *A* and *B*, or *C* and *D*, even though there is a difference in specified reach groups. Each colored line segment represents an individual reach.

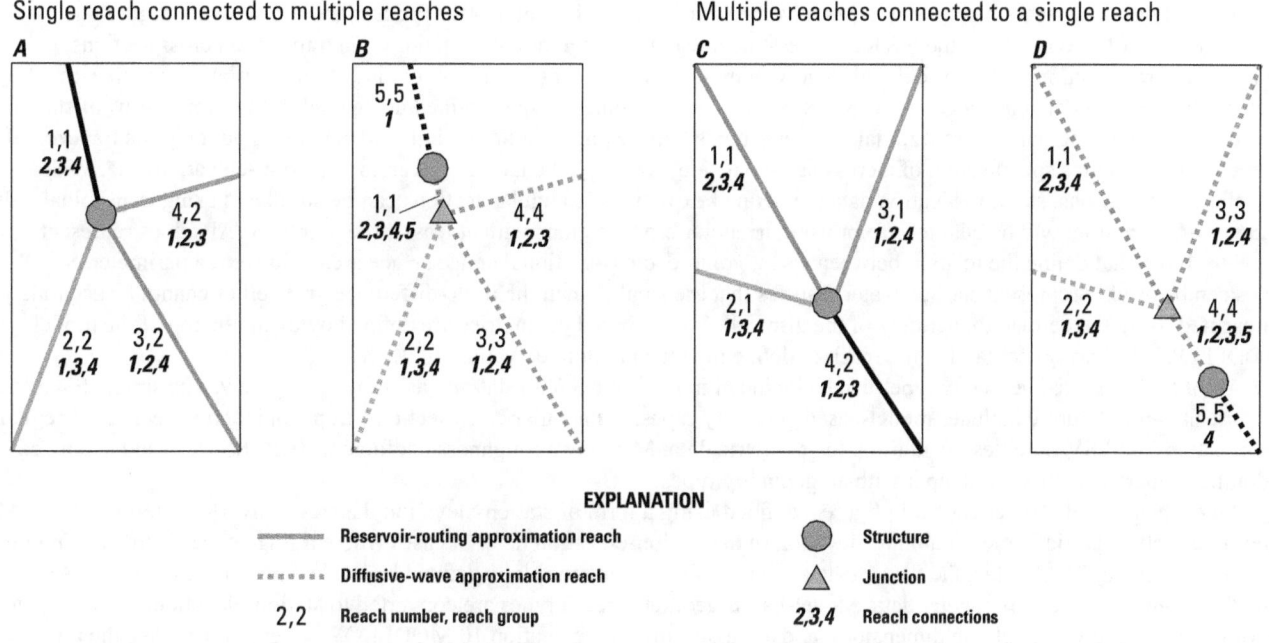

Figure 7. Examples of the simplified connectivity that should be developed for problems where a structure controls flow between, *A*, a single reservoir-routing approximation reach and multiple reservoir-routing approximation reaches, *B*, a single diffusive-wave approximation reach and multiple diffusive-wave approximation reaches, *C*, multiple reservoir-routing approximation reaches and a single reservoir-routing approximation reach, or *D*, multiple diffusive-wave approximation reaches and a single diffusive-wave approximation reach. Different color line segments represent individual reaches.

Structures that are part of a collection of reaches using the reservoir-routing approximation can easily be represented in the SWR1 Process. For example, there is no difference between connections when a structure is present (fig. 7A and 7C), or when the connection is a simple junction (fig. 6A and 6C). When reaches simulated by the diffusive-wave approximation are used, the presence of a structure can require modification of the network shown in figure 7A and 7C to correctly route structure flow at confluences. Because structure discharge is routed based on a single defined connection for a reach, it is necessary to split single reaches into two reaches, with one reach upstream and one reach downstream of the structure (fig. 7B and 7D). This splitting allows structure flow to be correctly routed prior to, or after, a confluence. For example, if the network shown in figure 7A was used with diffusive-wave approximation routing reaches and the structure was specified to be connected to the upstream end of reach 4, structure discharge would be routed to reach 4 and would not be correctly distributed to reaches 2, 3, and 4. Excess structure discharge routed to reach 4 would have to backflow to reach 2 and 3 in order for structure flow to be distributed to those reaches. This scenario is therefore not a physically realistic representation of the system. If the network shown in figure 7B were used, structure discharge from reach 5 to reach 1 would be correctly split between reaches 2, 3, and 4, based on hydraulic properties and simulated stage gradients. The numbering of reaches can be important. In cases where direct solution methods are used, and where the number of reach groups is large (greater than 1,000, for example), it can be advantageous in terms of required computer memory to simulate the problem to minimize the matrix bandwidth. For this application, the matrix bandwidth is equal to the largest numerical difference of reach group numbers for a connected set of reach groups. The matrix bandwidth does not affect performance when the surface-water system is solved iteratively. The direct and iterative methods available in the SWR1 Process are described in detail in appendix 3.

One of the most complicated aspects of creating a SWR1 dataset is intersecting one-dimensional polyline segments representing rivers, creeks, and canals with a MODFLOW grid. This intersection is required to create SWR1 reaches, reach connectivity, and reach groups. To facilitate this process, a simple pre-processor (SWRPre) has been developed that uses an Environmental Systems Research Institute (Esri) polyline shapefile that defines one-dimensional surface-water features and a MODFLOW discretization (DIS) file to create a data file that defines SWR1 reaches, reach groups, and reach connectivity. SWRPre also creates (1) an Esri polyline shapefile that contains the data contained in the SWR1 data file to validate the SWR1 network, and (2) an Esri polygon shapefile of the DIS file to validate the coordinate offset and rotation angle provided to the pre-processor. The input-data requirements and information on use of SWRPre are given in appendix 4.

Cross-Section Geometry Options

A variety of cross-section geometry types have been implemented in the SWR1 Process and are shown in figure 8. Geometry types (IGEOTYPE) in the SWR1 Process include (1) rectangular cross sections, (2) trapezoidal cross sections, (3) irregular cross sections, (4) user-defined stage-volume-area-perimeter (SVAP) tables, and (5) model-based geometries. Rectangular, trapezoidal, and irregular cross-sections represent geometry types defined using slightly different forms of station-elevation data. Rectangular and trapezoidal geometry types both require a bottom width and elevation, but only the trapezoidal geometry type requires specification of a cross-sectional side slope; these data are converted to station-elevation data. With irregular cross-sections, station-elevation data are defined explicitly. Triangular sections can be simulated using trapezoidal cross sections with a bottom width equal to zero or using irregular cross sections with at least three points. SVAP tables represent geometry types that define the relation between stage, volume, cross-sectional area, surface area, and wetted perimeter. SVAP tables can be used to represent surface-water features that are smaller than the finite-difference grid cell or cannot be accurately represented using the defined topography of the grid cell. Model-based geometries are defined by using the top of the model grid (MODFLOW Top array) containing the reach to define the reach bottom elevation (GBELEV).

All of the supported geometry types can be included in the same SWR1 dataset, as shown in figure 9. The simple SWR1 network shown in figure 9 includes model-based geometry types, rectangular cross sections, trapezoidal cross sections, irregular cross sections, and SVAP tables. In addition to geometric data, Manning's roughness coefficients (GMANNING) and reach bed hydraulic properties are required input with all geometry types.

All of the geometry types shown in figure 8 defined using a form of station-elevation data are converted internally to SVAP tables that include entries for each unique elevation in the station-elevation data. In cases where a stage exceeds the maximum stage in the internal SVAP table, the SVAP data are extrapolated using the slope defined by the last two entries in the SVAP table for volume, area, and perimeter data. Model-based-geometry reach types are converted to station-elevation data using the top of the model, row and column dimensions, and a maximum water elevation 10 MODFLOW length units above the top of the model (MODFLOW Top array). Finite-difference trapezoidal integration and reach length are used to calculate the relation between stage and volume, cross-sectional area, wetted perimeter, and surface area for each reach defined using a form of station-elevation data. The approach used to calculate reach SVAP tables from raw cross-section data is shown in figure 10. The approach to calculate reach SVAP tables results in an exact integration of the station-elevation data. If necessary, additional SVAP table values are calculated 1×10^{-6} and 10 MODFLOW length units above the reach bottom and the top of the model grid (MODFLOW Top array), respectively, to define the minimum and maximum non-zero volume, cross-sectional area, surface

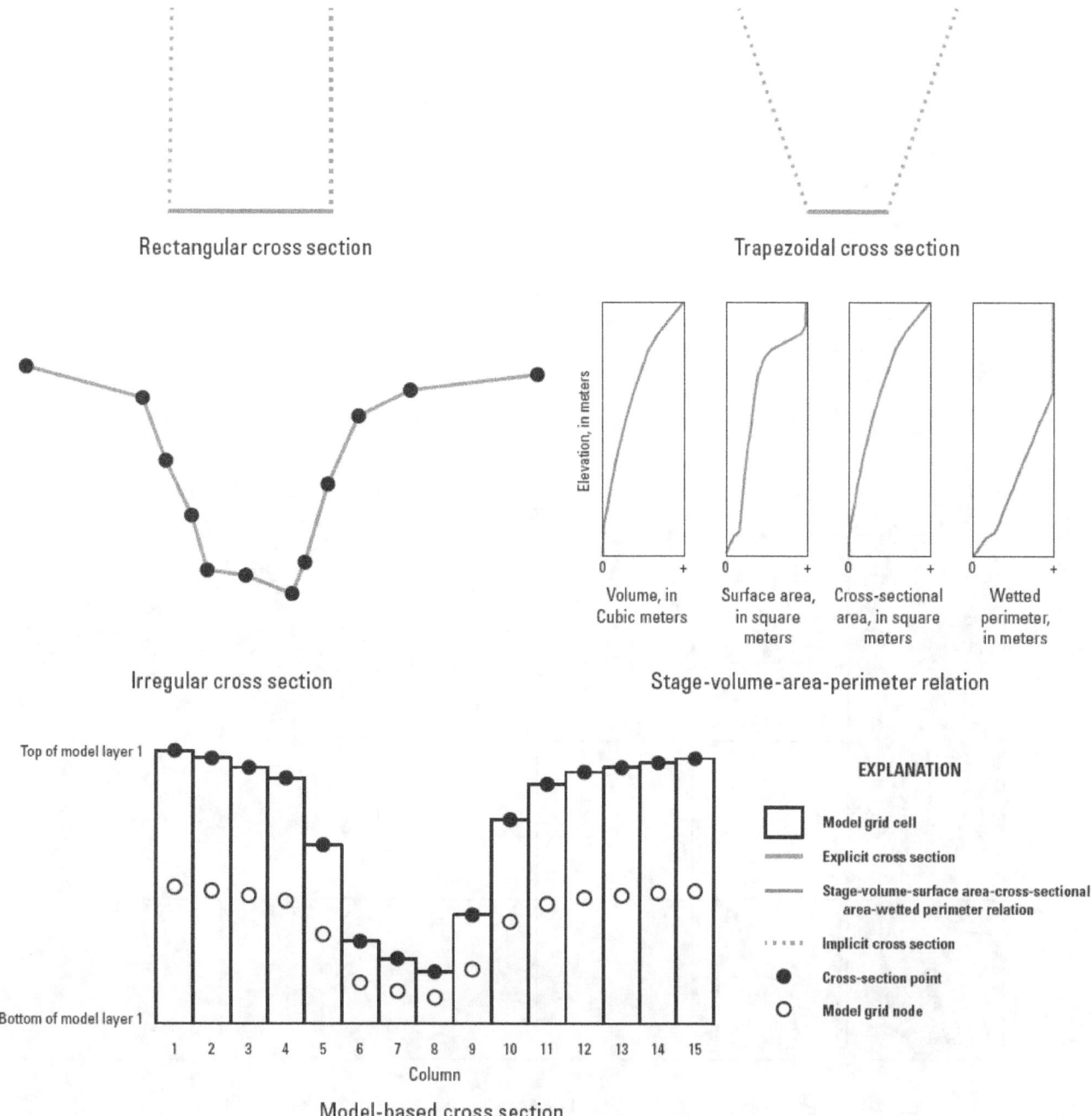

Figure 8. Geometry types supported by the SWR1 Process. The stage-volume-area-perimeter (SVAP) relation is representative of the irregular cross section.

area, and wetted-perimeter SVAP table entries. Intermediate SVAP table values are interpolated between the reach bottom and the top of the model (MODFLOW Top array) in the model row and column containing the reach. In the SWR1 Process, it is assumed that the MODFLOW Top array represents land surface. Intermediate SVAP table values are not interpolated for user-defined SVAP tables.

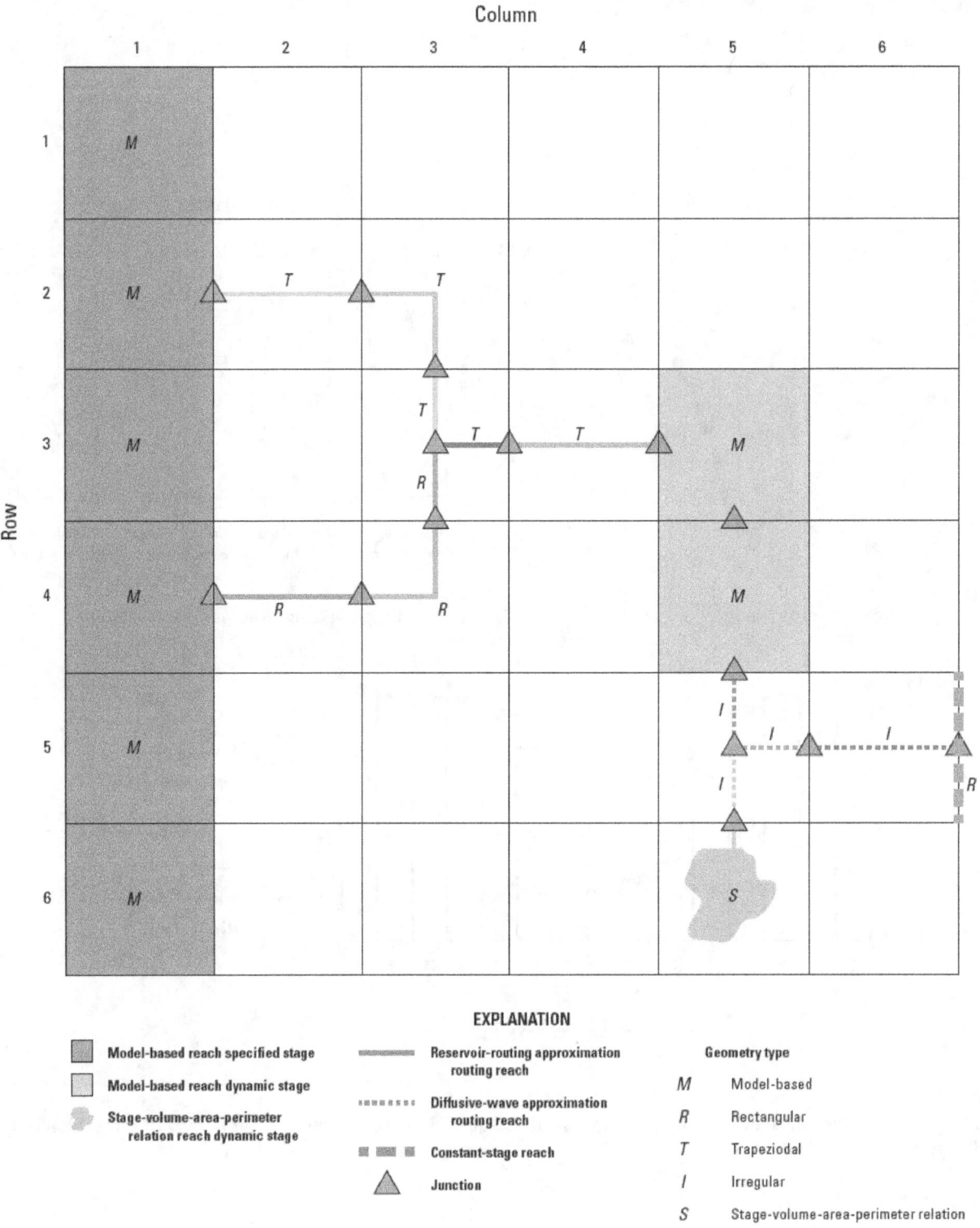

Figure 9.   Simple SWR1 network that includes multiple geometry types. This simple network also includes reservoir-routing reaches, reaches simulated by the diffusive-wave approximation, and constant-stage reaches. Different color line segments represent individual reaches.

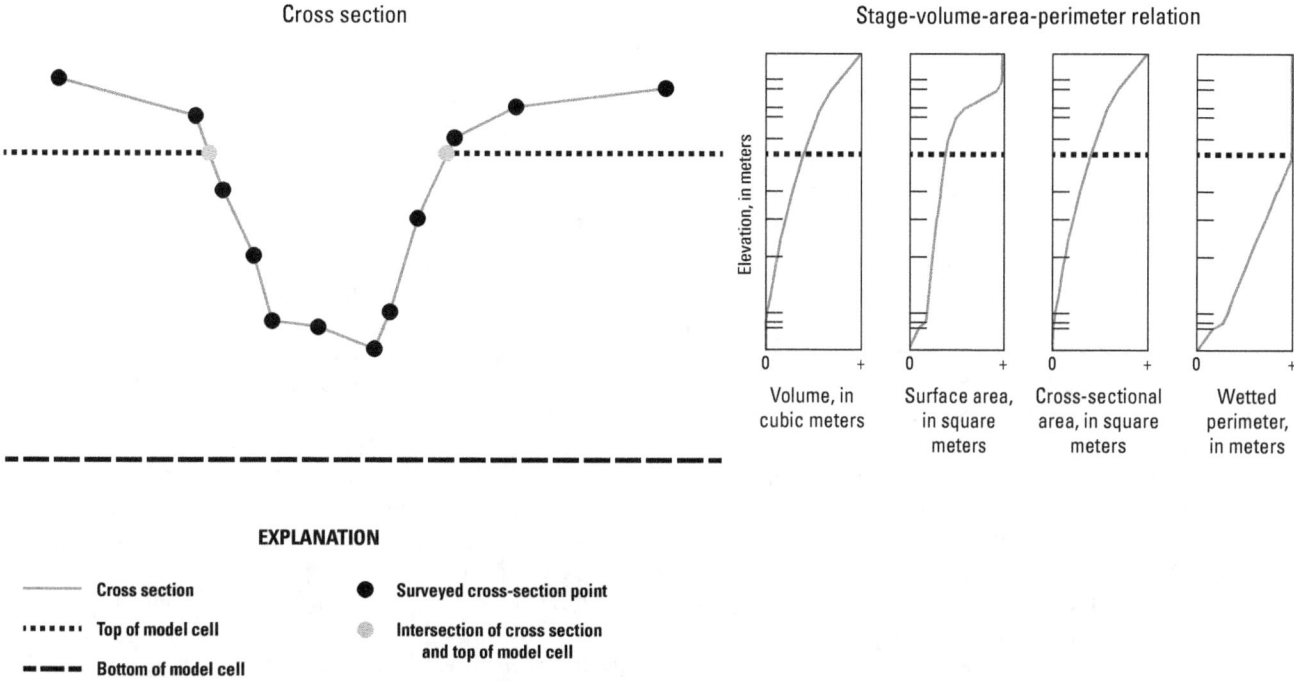

Cross section   Stage-volume-area-perimeter relation

Figure 10.   Example of the approach used to develop reach stage-volume-area-perimeter tables from raw geometric data and model layer elevation data.

Flow Terms for Reach Groups

This section details the discretized flow terms calculated for each reach group. As noted earlier, these flow terms are shown graphically in figure 3.

Reservoir-Routing and Diffusive-Wave Approximation Terms

With the reservoir-routing approximation, flow from one reach group to another (Q_M) can only occur through control structures (described in the next section). For flow between connected reaches simulated by the diffusive-wave approximation (eq. 9), the central-in-space finite-difference form of the $Q_M^{t+\Delta t}$ term at the current time in equation 17 is calculated using

$$Q_M^{t+\Delta t} = \text{sign}\left(\frac{\Delta h_{n-m}}{\Delta x_{n-m}}\right)\frac{c_M}{\overline{n}_{n-m}}\,\overline{A}_{n-m}\overline{R}_{n-m}^{2/3}\left(\left|\frac{\Delta h_{n-m}}{\Delta x_{n-m}}\right|\right)^{1/2},$$

(18)

where

Δh_{n-m} is the stage difference from reach group n to m,

Δx_{n-m} is the total distance from the center of reach group n to m,

$\overline{n}_{n-m}$ is the distance-weighted Manning's roughness coefficient [T/L$^{1/3}$] between reach group n to m based on defined roughness coefficients for reach group n to m,

$\overline{A}_{n-m}$ is the distance-weighted cross-sectional area [L^2] between reach group n to m, and

$\overline{R}_{n-m}$ is the distance-weighted hydraulic radius [L] between reach group n to m.

Surface-Water Control Structures

For reaches with control structures, anywhere from one to all of the Q_M terms in equation 17 are defined by structure equations. Multiple control structures can be defined within a single reach, and are treated as composite structures for which the total control structure flow is equal to the sum of the individual components. The total control structure flow in an arbitrary reach ii is

$$Q_{M_{ii}} = \sum_{is=1}^{ns_{ii}} Q_{S_{is}},$$

(19)

where

ns_{ii} is the number of control structures at connection ii in an arbitrary reach, and

$Q_{S_{is}}$ is the calculated flow [L³/T] for control structure is in reach ii.

A typical control structure that can be simulated with the SWR1 Process is shown in cross section along a reach length (fig. 11). For discussion purposes, this particular control structure can be considered an operable gated spillway. The control structure is a barrier to channel flow that spans the entire width of the stream or canal. An immovable part of the structure called the spillway is constructed in the bottom of the reach (normally with concrete) and has an invert elevation h_S. There is no flow through the structure when stage on both sides of the structure is less than h_S. Operable gates are located above the invert elevation, and can open and close to control flow between reach ii and reach jj. In this case, flow can occur in both directions and is determined by stage on both sides of the structure (h_{ii} and h_{jj}) and the resulting stage difference (Δh_{ii-jj}). The cross-sectional area of flow perpendicular to the flow direction is a function of the stream width, h_{ii}, h_{jj}, and h_S. The structure type (S_{TYPE}), controls the specific form of the equation used to calculate structure flow.

Operation of the gate can be simulated by the SWR1 Process to meet a target structure control stage or flow, v_{SC} (STRCRIT). For example, if the management objective is to maintain the stage in reach ii (h_{ii}) at an elevation less than v_{SC}, then the model will open and close the gate as necessary to try and meet this objective. A logical operator (L_{OPR}) specifies the desired relation (greater than or equal to, or less than) between reach stage or flow and v_{SC}. For some structures, flow may be limited by a maximum structure discharge (Q_{MAX}; STRMAX).

In cases where stage is above h_S on one side of the structure only, the stage difference (Δh) is calculated using the maximum stage (h_{max}) on either side of the structure (h_{ii} or h_{jj}) and h_S. Structures can be further simplified by assuming that there is no tailwater control, which results in unidirectional flow controlled by the reach stage (h_{ii}) and h_S. In some cases, high tailwater conditions can reduce structure flows. The relation between tailwater stage and the structure flow condition is shown in figure 12.

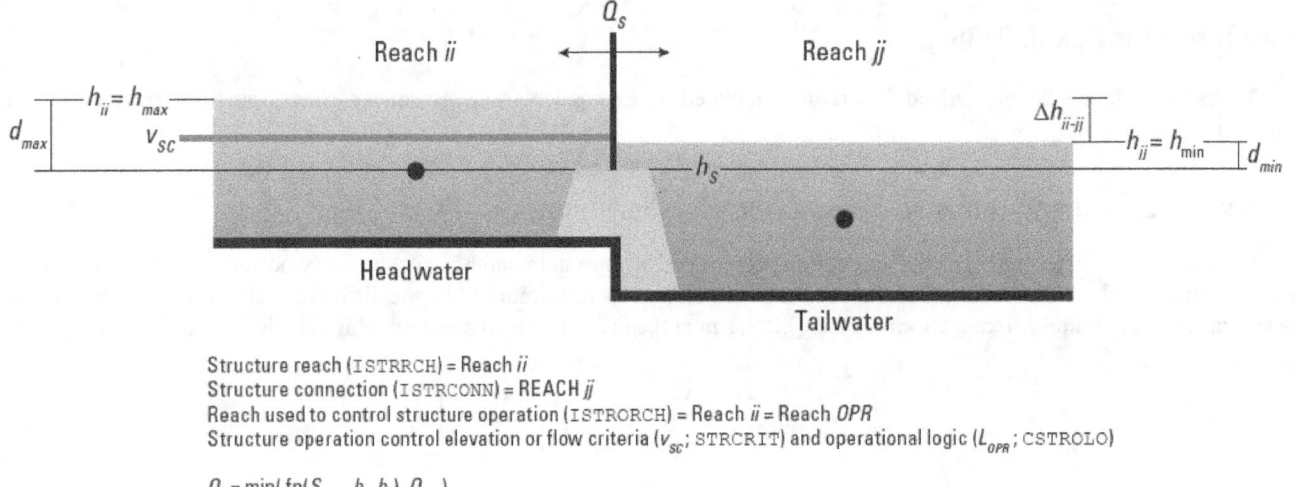

Structure reach (ISTRRCH) = Reach ii
Structure connection (ISTRCONN) = REACH jj
Reach used to control structure operation (ISTRORCH) = Reach ii = Reach OPR
Structure operation control elevation or flow criteria (v_{SC}; STRCRIT) and operational logic (L_{OPR}; CSTROLO)

$Q_S = \min(\ fn(S_{TYPE}, h_{ii}, h_{jj}), Q_{max})$

EXPLANATION

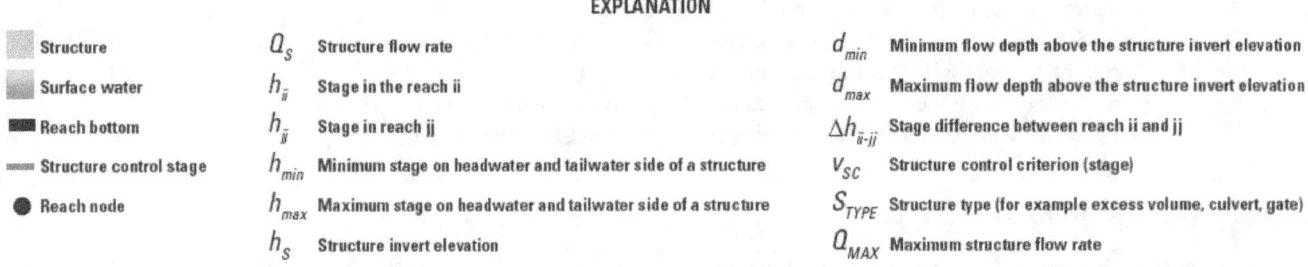

▨ Structure	Q_S Structure flow rate	d_{min} Minimum flow depth above the structure invert elevation
▨ Surface water	h_{ii} Stage in the reach ii	d_{max} Maximum flow depth above the structure invert elevation
▬ Reach bottom	h_{jj} Stage in reach jj	Δh_{ii-jj} Stage difference between reach ii and jj
═ Structure control stage	h_{min} Minimum stage on headwater and tailwater side of a structure	v_{SC} Structure control criterion (stage)
● Reach node	h_{max} Maximum stage on headwater and tailwater side of a structure	S_{TYPE} Structure type (for example excess volume, culvert, gate)
	h_S Structure invert elevation	Q_{MAX} Maximum structure flow rate

Figure 11.   Typical SWR1 Process control structure and static and dynamic parameters that affect structure discharge.

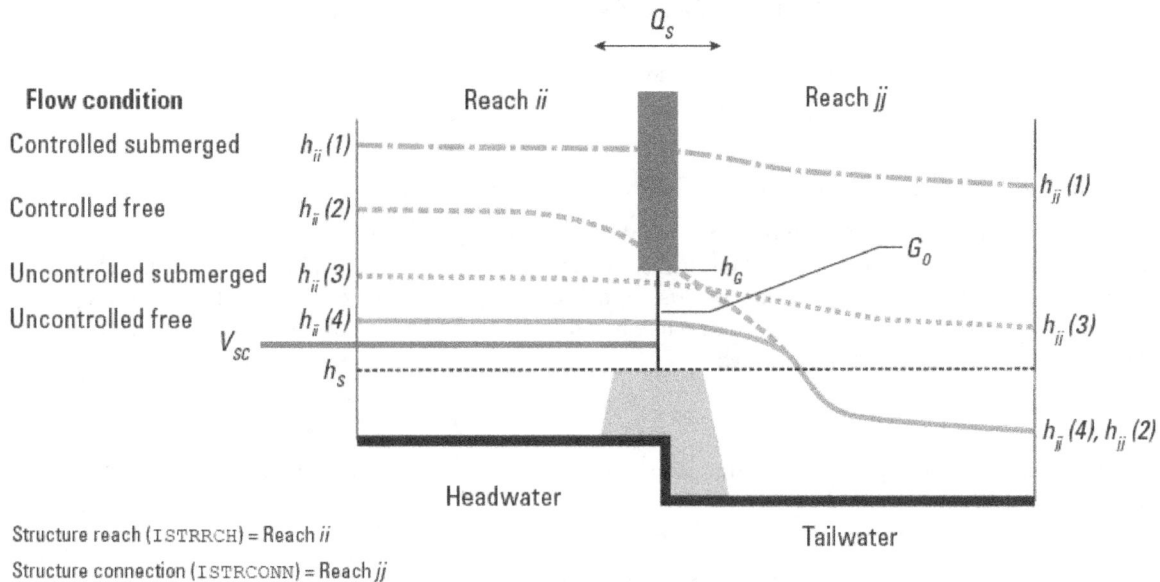

Flow condition

Controlled submerged

Controlled free

Uncontrolled submerged

Uncontrolled free

Structure reach (ISTRRCH) = Reach *ii*

Structure connection (ISTRCONN) = Reach *jj*

Reach used to control structure operation (ISTRORCH) = Reach *ii* = Reach *OPR*

EXPLANATION

▨	Structure	▦	Reach stage – flow condition 2	Q_s	Structure flow rate	h_s Structure invert elevation
▮	Gate	▦	Reach stage – flow condition 3	h_{ii}	Stage in reach *ii*	h_G Structure gate elevation
▬	Reach bottom	▦	Reach stage – flow condition 4	h_{jj}	Stage in reach *jj*	G_O Gate opening
▦	Reach stage – flow condition 1	▦	Structure control stage	V_{SC}	Structure control criterion (stage)	

Figure 12. Flow regimes for control structures subject to tailwater conditions (adapted from figure 9 of Swain and others, 1996). Flow regime numbering is based on the flow classification defined by Otero (1995).

For structures affected by tailwater conditions, transition of flow conditions between free and submerged conditions is controlled by (1) the ratio of the minimum flow depth above the structure invert ($d_{\min} = \max\left(0, h_{\min} - h_s\right)$) and the maximum flow depth above the structure invert ($d_{\max} = \max\left(0, h_{\max} - h_s\right)$) and for gated spillways; and (2) the ratio of the $d_{\min}$ to the gate opening, G_O. The gate elevation, h_G, relative to $h_{\max}$ or other vertical obstruction, such as the top of a culvert, determines if flow conditions transition from uncontrolled to controlled conditions. The relation of flow conditions to $d_{\min}$ to G_O and $h_{\min}$ to $h_{\max}$ is summarized next (Ansar and Chen, 2009; Barkau, 1996; Chaudhry, 2008).

Controlled submerged-flow conditions (flow condition 1) occur for gated spillways or culverts when

$$\frac{d_{\min}}{G_O} \geq \begin{cases} 1.0 \text{ for gated spillways} \\ 1.2 \text{ for culverts} \end{cases}, \; h_{\max} \geq h_G, \text{ and } \frac{d_{\min}}{d_{\max}} > 0. \tag{20}$$

Controlled free-flow conditions (flow condition 2) occur for gated spillways or culverts when

$$\frac{d_{\min}}{G_O} \geq \begin{cases} 1.0 \text{ for gated spillways} \\ 1.2 \text{ for culverts} \end{cases}, \; h_{\max} \geq h_G, \text{ and } \frac{d_{\min}}{d_{\max}} = 0. \tag{21}$$

Uncontrolled submerged-flow conditions (flow condition 3) occur for gated spillways, culverts, or weirs when

$$\frac{d_{\min}}{G_O} < \begin{cases} 1.0 \text{ for gated spillways} \\ 1.2 \text{ for culverts} \end{cases}, \; h_{\max} < h_G, \text{ and } \frac{d_{\min}}{d_{\max}} > 0. \tag{22}$$

Uncontrolled free-flow conditions (flow condition 4) occur for gated spillways, weirs, or culverts when

$$\frac{d_{\min}}{G_O} < \begin{cases} 1.0 \text{ for gated spillways} \\ 1.2 \text{ for culverts} \end{cases}, \ h_{\max} < h_G, \text{ and } \frac{d_{\min}}{d_{\max}} = 0.$$

(23)

No structure flow occurs if $h_{\max}$ is less than h_S. For weirs, the $\frac{d_{\min}}{G_O}$ condition is not evaluated.

The control structures included in the SWR1 Process represent a number of typical structures (ISTRTYPE) and include (1) a simple excess-volume structure, (2) a uncontrolled discharge connection structure controlled by bed roughness and reach stage gradients, (3) a pump, (4) circular and rectangular culverts, (5) a fixed crest weir, (6) a fixed-gate spillway, (7) an operable movable crest weir, and (8) an operable gated spillway. A structure based on a user-defined rating curve has also been implemented and can be used to define custom control structures.

Excess Volume

An excess-volume structure discharges water when the stage in reach *ii* exceeds a specified control elevation. Excess-volume structure discharge is calculated using

$$Q_S = \begin{cases} \min\left(Q_{\max}, \dfrac{V_{v_{SC}}}{\Delta t} \right) & (h_{ii} > v_{SC}), \\ 0 & (h_{ii} \le v_{SC}) \end{cases}$$

(24)

where

$V_{v_{SC}}$ is the available volume [L³] above the control elevation on the upstream side of the structure (reach *ii*).

Excess-volume structure discharge is the minimum of the maximum flow rate for the structure and the available volume above the control elevation when the control elevation is exceeded; as a result, it is possible to discharge the entire excess volume in a single SWR1 time step if the product of $Q_{\max}$ and Δt exceeds $V_{v_{SC}}$.

Uncontrolled Discharge Connection

In order to allow for uncontrolled discharge between reaches to be simulated using the reservoir-routing approximation, an uncontrolled discharge connection based on the diffusive-wave approximation was implemented. The central-in-space finite-difference form of the diffusive-wave approximation of the Saint-Venant equations used for uncontrolled-gradient connections is

$$Q_S = \text{sign}\left(\frac{\Delta h_{jj-ii}}{\Delta x_{jj-ii}} \right) \frac{c_M}{\overline{n}_{jj-ii}} \overline{A}_{jj-ii} \overline{R}_{jj-ii}^{2/3} \left(\left| \frac{\Delta h_{jj-ii}}{\Delta x_{jj-ii}} \right| \right)^{1/2},$$

(25)

where

Δh_{jj-ii} is the stage difference over the structure from reach *jj* to *ii*,

Δx_{jj-ii} is the total distance from the center of reach *jj* to *ii*,

$\overline{n}_{jj-ii}$ is the distance-weighted Manning's roughness coefficient [T/L^{1/3}] between reach *jj* and *ii* based on defined roughness coefficients for reach *ii* and *jj*,

$\overline{A}_{jj-ii}$ is the distance-weighted cross-sectional area [L²] between reach *jj* to *ii*, and

$\overline{R}_{jj-ii}$ is the distance-weighted hydraulic radius [L] between reach *jj* and *ii*.

Equation 25 is the finite difference form of the diffusive-wave approximation of the Saint-Venant equations (eq. 9). In cases where h_{ii} is greater than h_{jj}, $\text{sign}\left(\dfrac{\Delta h_{jj-ii}}{\Delta x_{jj-ii}} \right)$ is negative and flow is from reach *ii* to reach *jj*. Conversely, when h_{jj} is greater than h_{ii}, $\text{sign}\left(\dfrac{\Delta h_{jj-ii}}{\Delta x_{jj-ii}} \right)$ is positive and flow is from reach *jj* to reach *ii*.

This structure type also can be used to define zero- or critical-depth gradient boundary conditions for reaches (undefined structure connection – ISTRCONN). For zero-depth gradient boundaries, $\overline{n}, \overline{A},$ and $\overline{R}$ from reach *ii* are used; the reach bottom slope, S_o, for zero-depth gradient boundaries is calculated as the difference between the simulated stage in reach *ii* and the sum of a user-defined bottom elevation (STRINV) at the structure and the water depth in reach *ii* over half the length of reach *ii*. In cases where a user-defined bottom elevation at the structure is not specified, a critical-depth boundary condition is applied. The critical-depth boundary condition defines the depth at the boundary to be the critical depth under free-fall conditions and is expressed as

$$Q_S = \left(gA^2 R\right)^{1/2}.$$

(26)

Pump

The simplest control structure implemented in the SWR1 Process is the pump (specified-discharge structure). For this structure, flow is calculated as

$$Q_S = \begin{cases} Q_{MAX} & \left(\text{if } v_{OPR} \text{ satisfies } L_{OPR} \text{ constraint for } v_{SC}\right) \\ 0 & \left(\text{otherwise}\right) \end{cases},$$

(27)

where v_{OPR} is the simulated stage or flow in the user-defined reach *OPR*. The L_{OPR} term can be user-defined as less than v_{SC}, or greater than or equal to v_{SC}.

As implemented, a pump is a unidirectional structure that discharges water from reach *ii* to reach *jj*, connected to the end of reach *ii*, when the simulated stage or flow in reach *OPR* violates the user-defined logic relative to the user-defined control elevation v_{OPR}. For example, assigning the control reach *OPR* to be reach *ii* and setting the user-defined control elevation or flow, v_{OPR}, at or below the bottom of reach *ii* (GBELEV) or zero, respectively, will define Q_S to be Q_{max} for all non-zero depths or flows in reach *ii*. A pump can be used to represent a physical pump or to fix structure flows at specified rates. For example, a pump can be used during the calibration phase of a modeling project, to set structure flow and modify model parameters to balance terms in the surface-water budget.

Fixed Control Structures

Fixed geometry control structures included in the SWR1 Process include circular and rectangular culverts, fixed crest weirs, and fixed-gate spillways. All fixed control structures can account for free and submerged flow conditions.

Culverts

When the ratio of the maximum water depth above the culvert invert ($d_{max} = h_{max} - h_S$) to the diameter of a circular culvert or rise (height) of a rectangular culvert is greater than 1.2, and h_{min} is below the top of the culvert on the downstream side, the entrance is considered submerged (Chaudhry, 2008). Submerged entrance culvert flow is calculated using the orifice equation

$$Q_S = C_C A_C \left(2g\left(d_{max} - C_C D_C\right)\right)^{1/2},$$

(28)

where
$\quad C_C \quad$ is the culvert discharge coefficient [unitless] (STRWCD),
$\quad A_C \quad$ is the cross-sectional area of the culvert [L²], and
$\quad D_C \quad$ is the diameter of the circular culvert (STRWID) or the rise of the rectangular culvert [L] (STRWID2).

For sharp- and round-edge inlets, the culvert discharge coefficient, C_C, is typically specified to be 0.6 and 0.8, respectively, when specific data are not available for a structure (Chaudhry, 2008).

When the culvert entrance is submerged and h_{min} is above the top of the culvert on the downstream side, the culvert is outlet controlled and friction losses in the culvert are considered. Culvert flow under outlet control is calculated using

$$Q_S = \frac{A_C\left(2g\left(h_{max} - h_{min}\right)\right)^{1/2}}{\sqrt{1 + C_C + \dfrac{2gn_C^2 l_C}{c_M^2 R_C^{4/3}}}},$$

(29)

where

n_C is the Manning roughness coefficient of the culvert [$T/L^{1/3}$] (STRMAN),
l_C is the culvert length parallel to flow [L] (STRLEN), and
R_C is the hydraulic radius of the culvert [L].

When d_{max} is less than 1.2 times the diameter of a circular culvert or the rise of a rectangular culvert, the entrance is considered unsubmerged (Chaudhry, 2008). Unsubmerged culvert flow is calculated using

$$Q_S = C_C A_C \left(2g d_{max}\right)^{1/2}.$$

(30)

The relations between the hydraulic depth (d_{hC}), radius (r_C), and area (A_C) of the culvert for rectangular culverts are shown in figure 13, and defined using

$$R_C = \min\left(d_{max}, C_{RISE}\right) \text{ and}$$

(31)

$$A_C = C_{SPAN} R_C,$$

(32)

where

C_{RISE} is the culvert rise [L] (STRWID2) and
C_{SPAN} is the culvert span [L] (STRWID).

The hydraulic depth, d_{hC}, is defined as

$$d_{hC} = \min\left(h_{max} - z_{max}, C_{RISE}\right),$$

(33)

where z_{max} is the invert elevation of the culvert on upstream side (h_{max} side) [L].

For circular culverts, the relation between d_{hC}, h_{max}, R_C, and A_C is more complicated. The variables used to calculate R_C and A_C for circular culverts are shown in figure 13.

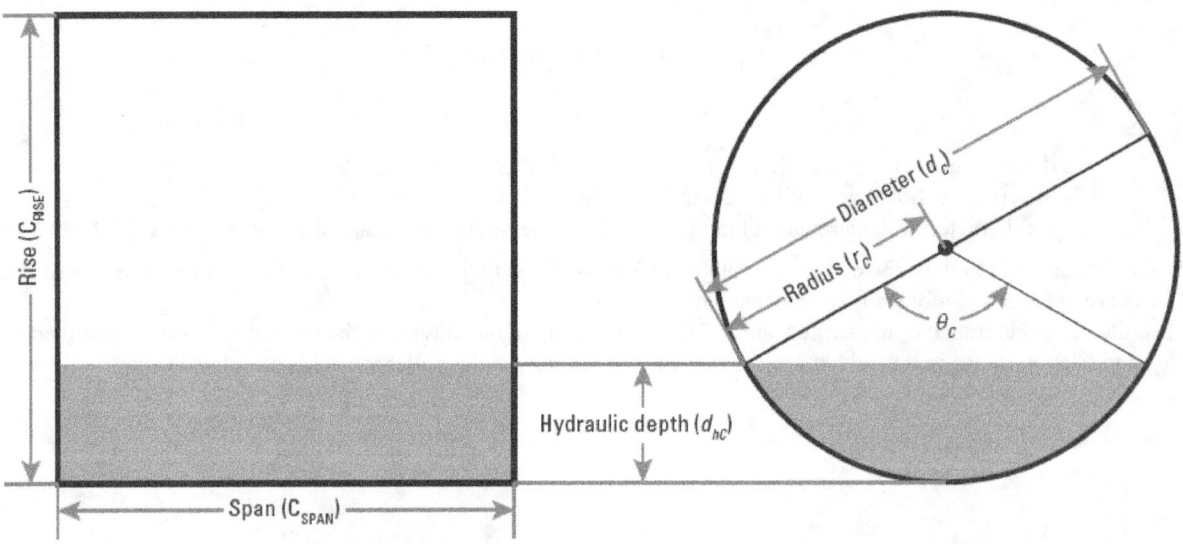

Figure 13.   Variables used to calculate the hydraulic radius and flow area for partially-filled culverts.

If d_{hC} is less than the culvert radius, r_C, then the angle, θ_C, between the center of the culvert and the top of the water surface in the culvert is calculated as (French, 1985)

$$\theta_C = \begin{cases} 0 & (d_{hC} = 0) \\ 2\cos^{-1}\left(1 - \dfrac{2d_{hC}}{d_C}\right) & (0 < d_{hC} < d_C) \\ 2\pi & (d_{hC} = d_C) \end{cases} ,$$

(34)

where d_C is the diameter of the culvert [L]. R_C is calculated as (French, 1985)

$$R_C = \frac{d_C}{4}\left(1 - \frac{\sin\theta_C}{\theta_C}\right).$$

(35)

A_C is calculated as (French, 1985)

$$A_C = \frac{d_C^2}{8}\left(\theta_C - \sin\theta_C\right).$$

(36)

Fixed Crest Weirs

The generic weir equation of Barkau (1996) can transition between free and submerged flow conditions, and has been implemented in the SWR1 Process in a modified form. The generic weir equation is

$$Q_S = C_D C_F W_S d_{max}\left(2g d_{max}\right)^{1/2},$$

(37)

where

C_D is the weir discharge coefficient [unitless] (STRWCD),
C_F is a submergence factor [unitless], and
W_S is the weir length perpendicular to flow [L] (STRWID).

C_F is defined as

$$C_F = \begin{cases} \left[1 - \dfrac{d_{min}}{d_{max}}\right]^{c_3} & \left(\dfrac{d_{min}}{d_{max}} > 0\right) \\ 1 & \left(\dfrac{d_{min}}{d_{max}} = 0\right) \end{cases} ,$$

(38)

where c_3 (STRCD3) is a submergence exponent [unitless] and can be specified to be 0.5 when specific data are not available for a structure.

In other surface-water routing codes, a dimensional weir discharge coefficient, C_w [$L^{1/2}$/T], is commonly specified (for example, HEC-RAS; U.S. Army Corps of Engineers, 2008). C_w is defined as

$$C_W = C_D\left(2g\right)^{1/2}.$$

(39)

When C_w is used, equation 37 is redefined as

$$Q_S = C_W C_F W_S d_{max}\left(d_{max}\right)^{1/2}.$$

(40)

In the SWR1 Process, C_D should be specified, and dimensional C_W values from existing models or compiled tables should be multiplied by $\left(2g\right)^{-1/2}$.

Fixed-Gate Spillway

The equations used for a fixed-gate spillway are based on a combination of the generic weir equation (eq. 37) and a form of the orifice equation (eq. 28) that smoothly transitions between unsubmerged and submerged flow conditions. The flow conditions for a gated spillway are shown in figure 12. Controlled free-flow conditions occur when G_O is less than d_{max}, $G_O \leq 0.8\, d_{max}$, and d_{min} is zero. The orifice equation for fixed gate spillways is

$$Q_S = C_O C_F W_S G_O \left(2gd_{max}\right)^{1/2}.$$

(41)

where C_O (STRWCD2) is the orifice discharge coefficient [unitless]. C_O values of 0.61 and 0.8 are reasonable values for sharp and round-edge inlets, respectively, when a specific orifice discharge coefficient is not available for a structure (Chaudhry, 2008).

When G_O is greater than d_{max}, the gate does not affect the flow and equation 37 is used. When G_O is less than d_{max} and $G_O > 0.8\, d_{max}$, equation 41 is modified to use a transition discharge coefficient

$$C_T = C_O + \frac{C_D - C_O}{0.2 d_{max}}\left(G_O - 0.8 d_{max}\right),$$

(42)

where C_T is the transition discharge coefficient [unitless]. The revised form of equation 37 used for transitional flow conditions is

$$Q_S = C_T C_F W_S G_O \left(2gd_{max}\right)^{1/2}.$$

(43)

Operable Control Structures

An operable pump, a movable crest weir, and an operable gated spillway have been included in the SWR1 Process. This addition allows the structure discharge, Q_S, weir crest, h_S, and gate elevation, h_G, for operable pumps, weirs, and gated spillways, respectively, to vary from one SWR1 time step to another based on simulated stages or flows.

Pump

Pumps are used to maintain stages at specified values and prevent over-drainage of reaches upstream of the structure. The operable pump is turned on when the operational criteria for the structure are met. To prevent over-drainage, the pump is turned off when the operational criteria for the structure are not met.

Movable Crest Weir

The equation used to calculate flow for the movable crest weir is identical to the fixed crest weir equation. A movable crest weir allows flow anytime h_{max} is greater than the top of the structure. This type of structure is often used to maintain stages at specified levels upstream of the structure. The movable crest weir is opened when the operational criteria for the structure are met to allow additional discharge through the structure. To prevent over-drainage, the movable crest weir is closed when the operational criteria for the structure are not met.

Gated Spillway

The equation used to calculate flow for the operable gated spillway is identical to the fixed-gate spillway equation. Gated spillways are also used to maintain stages at specified levels and prevent over-drainage of reaches upstream of the structure. The operable gated spillway is opened when the operational criteria for the structure are met to allow additional discharge through the structure. To prevent over-drainage, the gated spillway is closed when the operational criteria for the structure are not met.

Structure Operations and Control Locations

Operation of the movable crest weir and gated spillway are similar, but the elevation of the weir crest is decreased and the elevation of spillway gate is increased when the structure is opened to allow increased discharge from the structure. Conversely, the elevation of the weir crest is increased and the elevation of the spillway gate is decreased when the structure is closed to reduce discharge from the structure. Operation of a pump is identical to a gated spillway except the structure flow is adjusted instead of the structure gate elevation.

Movable crest weirs and spillway gates are closed and opened, respectively, using

$$h_G^{t+\Delta t} = \min\left(h_{G\max}, h_G^t + Rate_G \Delta t\right),$$

(44)

where

$h_G^{t+\Delta t}$ is the elevation of the structure during the current SWR1 time step [L],
$h_{G\max}$ is the maximum weir crest or spillway gate elevation [L] (STRMAX),
h_G^t is the structure elevation during the previous SWR1 time step [L], and
$Rate_G$ is the gate closing and opening rate [L/T] (STRRT).

In the opposite case, movable crest weirs and spillway gates are opened and closed, respectively, using

$$h_G^{t+\Delta t} = \max\left(h_S, h_G^t - Rate_G \Delta t\right).$$

(45)

Operable pumps are turned on and off using modified forms of equations 44 and 45 that substitute $Q_{S_{is}}^{t+\Delta t}$, $Q_{\max}$, and $Q_{S_{is}}^t$ for $h_G^{t+\Delta t}$, $h_{G\max}$, and h_G^t, respectively.

Structures can be operated when the stage is less or greater than (L_{OPR}) the specified control elevation or discharge (v_{SC}) at the reach controlling structure operations (OPR reach). The simplest case is when the reach connected at the structure is used as the control location (OPR reach = reach ii). However, specifying a reach other than the reach connected at the structure allows structures to be operated based on simulated stage or flow conditions elsewhere in the surface-water system. For example, the specified control elevation could be defined at a reach farther downstream from the connected reach (for example, reach ii + 5) with a control elevation that allows a structure to open when stages are less than a specified value. This structure operation criterion would allow downstream stage to be maintained at a specified elevation using water available upstream of the structure (for example, a reservoir). The SWR1 Process currently does not allow multiple control locations or multiple control elevations and (or) flows to be specified for individual control structures ($Q_{S_{is}}$).

Rating-Curve Structures

Rating-curve structures allow simple, user-defined structures to be defined in a SWR1 dataset. Structure flow is defined using a specified stage-discharge relation. Flow values are linearly interpolated between defined stage-discharge points. The stage-discharge slope between the last two entries is used to extrapolate stage-discharge values above maximum defined elevations. An example of a rating-curve structure representing the stage-discharge relation for a v-notched weir with an invert elevation of 0 ft, a weir side slope of 1.0 (v-notch angle = 90°), a discharge coefficient of 0.61, and a maximum water-surface elevation of 2 m is shown in table 3. The discharge equation for a v-notch weir is (Chaudhry, 2008)

$$Q_S = \frac{8}{15} C_D \tan\left(\frac{\theta_{VW}}{2}\right) d_{VW}^2 \left(2g d_{VW}\right)^{1/2},$$

(46)

where

θ_{VW} is the angle of the v-notch [degrees], and
d_{VW} is the flow depth of the v-notch weir [L].

Table 3. Example rating-curve structure representing a free-flowing v-notch weir.

Elevation, in meters	Discharge, in cubic meters per second	Elevation, in meters	Discharge, in cubic meters per second
0	0.00	1.2	2.27
.2	.03	1.4	3.34
.4	.15	1.6	4.67
.6	.40	1.8	6.26
.8	.82	2	8.15
1	1.44		

Specified Inflows and Outflows

Rainfall, evaporation, and lateral flows are assigned by reach, and the values can be modified by stress period. Lateral flows can represent point flows from within the model domain and external sources of water from outside the model domain. Rainfall is applied to the maximum reach perimeter of each reach; evaporation is removed from the wetted portion of each reach calculated using the current stage. Rainfall, evaporation, and lateral flows can be defined by reach using list format input or using two-dimensional arrays read using the MODFLOW U2DREL subroutine. If two-dimensional arrays are used, all reaches in the same row and column are given the same rainfall and evaporation rate. Lateral flows entered as two-dimensional arrays are distributed by the ratio of reach length to total sum of reach lengths in the same row and column. For example, if two reaches having lengths of 50 and 100 m are located within the same row and column, 33 and 67 percent of the lateral flow will be distributed to the reaches, respectively. Rainfall, evaporation, and lateral flow can also be modified at each SWR1 time step using optional external time series data files.

Aquifer-Reach Exchanges

SWR1 has been developed to allow coupling of reach segments to a single specified groundwater-model layer or automatic coupling of reach segments to multiple groundwater-model layers based on reach geometry data and MODFLOW layer elevations (fig. 2).

Aquifer-reach exchange in equation 15 is calculated for each reach using

$$Q_{AQ} = \sum_{k\ kl\min_{ii}}^{kl\max_{ii}} C_{ii,k}\left(H_{j,i,k} - h_{ii}\right),$$

(47)

where

$kl\min_{ii}$ is the minimum layer intersected by reach ii geometry,
$kl\max_{ii}$ is the maximum layer intersected by reach ii geometry,
$C_{ii,k}$ is a aquifer-reach conductance value for reach ii and layer k [L^2/T],
$H_{j,i,k}$ is the groundwater head for the cell that contains reach ii (column j, row i, and layer k) [L].

In the SWR1 Process, positive and negative Q_{AQ} values represent a source and sink for a reach, respectively. Aquifer-reach exchanges in MODFLOW are equal but opposite in sign to SWR1 Q_{AQ} values and consistent with the MODFLOW sign convention.

Options for Calculation of Reach Bed Conductance

Equation 47 represents the general equation used to calculate aquifer-reach exchanges. A number of different options (IGCNDOP) for calculating $C_{ii,k}$ have been implemented in SWR1. Options include specification of (1) a specified conductance that does not vary during a MODFLOW stress period, which is the approach used in the standard River Package in MODFLOW, and (2) a dynamic conductance that can vary by SWR1 time step and is calculated based on the maximum exchange perimeter and horizontal hydraulic conductivity and (or) hydraulic properties of reach-bed sediments. The specified conductance (GCND) is calculated as

$$C_{ii,k} = \frac{K_{b'}A_r}{b'},$$

(48)

where

$K_{b'}$ is the hydraulic conductivity of reach-bed sediments [L/T],
A_r is the surface area of the bottom of the reach [L^2], and
b' is the thickness of reach-bed sediments [L].

Conductance can be dynamically calculated based on reach-bed sediments and the maximum exchange perimeter by

$$C_{ii,k} = K'_{REACH}\chi_{p_k}l_{REACH},$$

(49)

where

K'_{REACH} is the leakance of reach-bed sediments [1/T] (GLK),
χ_{p_k} is the maximum exchange perimeter in layer k [L], and
l_{REACH} is the length of the reach [L] (RLEN).

The leakance of reach-bed sediments (K'_{REACH}) is calculated as K_b divided by b'. The maximum exchange perimeter in a layer is calculated from the stage-wetted perimeter relation for a reach as the difference between (1) the wetted perimeter calculated using the maximum of the reach stage, the groundwater level in the layer, and the bottom of the model layer for values less than the top of the model layer, and (2) the wetted perimeter calculated using the maximum of the bottom of the reach and the bottom of the model layer.

Alternatively, conductance can be calculated based on aquifer hydraulic conductivity using

$$C_{ii,k} = \frac{K_{H_{j,i,k}} \mathcal{X}_{P_k} l_{REACH}}{l_{R \to node}},$$ (50)

where

$K_{H_{j,i,k}}$ is the horizontal hydraulic conductivity of the model cell [L/T] and

$l_{R \to node}$ is the average horizontal distance from the reach to the center of the grid cell [L] (GCNDLN).

Conductance can also be calculated as a weighted combination of the leakance of reach-bed sediments and aquifer hydraulic conductivity using

$$C_{ii,k} = \frac{1}{\dfrac{l_{R \to node}}{K_{H_k} \mathcal{X}_{P_k} l_{REACH}} + \dfrac{1}{K'_{REACH} \mathcal{X}_{P_k} l_{REACH}}}.$$ (51)

The conductance options defined in equations 48 and 49 assume head loss between the reach and the aquifer occurs across b' and K', respectively; in equations 50 and 51, a portion or all of the head loss occurs over a defined interaction length within the aquifer. The average horizontal distance from the reach to the center of the grid cell, $l_{R \to node}$, should be greater than zero to prevent calculation of an infinite conductance.

Different conductance options can be used for different reaches in a SWR1 dataset and conductance options and parameters can change with succesive stress periods. For example, a natural reach could be initially simulated using the weighted combination of the leakance of reach-bed sediments and aquifer hydraulic conductivity option. Later in the simulation, the reach could be converted to a specified conductance option with a conductance of 0.0 to simulate installation of an impermeable lining to the reach. Another example would be a reach converted from a reach-bed conductance option to the aquifer hydraulic conductivity conductance option to represent canal modifications that expose highly conductive aquifer materials.

Vertical Distribution of Conductance

When the option to connect a reach with multiple groundwater-model layers is activated in one or more reaches, a composite reach conductance (eq. 47) is calculated based on the maximum exchange perimeter of the reach in each layer ($\mathcal{X}_{P_k}$). The general approach used to vertically distribute reach conductance is shown in figure 14. To vertically distribute conductance for each of the reach-bed conductance options, the length of exchange perimeter in each model layer is calculated and used as the exchange perimeter term in equations 49, 50, or 51. It is not possible to assign different conductance options to individual model layers connected to a single reach.

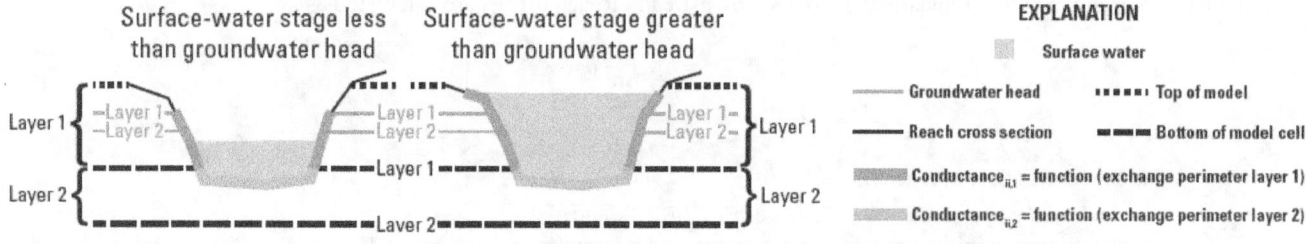

Figure 14.   Example of the approach used to calculate reach conductance for reaches that intersect multiple groundwater layers. For example, when the surface-water stage is lower than the groundwater head, the exchange perimeter in the layer is calculated using the groundwater head. Alternatively, when surface-water stage is higher than the groundwater head, the exchange perimeter in the layer is calculated using the surface-water stage.

For reaches that use the specified-conductance option, the specified conductance is distributed to individual layers based on the fraction of the total exchange perimeter each model layer intersects. This approach can be problematic if reach stages vary significantly during a simulation. In such cases, the fraction of the total exchange perimeter would vary with the reach stage. For example, in a model with two layers of equal thickness, the total specified conductance would be in layer 2 if stage were at the bottom of layer 1 and the model cell in layer 1 was dry. Alternatively, if stage were at the top of model layer 1, the specified conductance would be divided equally between model layers 1 and 2. There is no physical basis for such a change in reach conductance with stage, and would likely lead to non-convergence for some problems. As a result, it is recommended that reaches using the specified-conductance option be connected to a specific model layer.

To accurately simulate the exchange perimeter for aquifer-reach exchanges, the exchange perimeter used in equations 48, 49, 50, or 51 is calculated as the difference between the wetted perimeter calculated using (1) the higher of the reach stage and the groundwater head in the layer or the bottom of the overlying layer for layers other than the top layer; and (2) the higher of the reach bottom and the bottom of the model layer. For cases where the surface-water stage is lower than the groundwater head, the exchange perimeter in the layer is calculated using the groundwater head because the exchange perimeter extends from the water-table elevation to the bottom of the layer or bottom of the reach (fig. 14). Alternatively, when surface-water stage is higher than the groundwater head, the exchange perimeter in the layer is calculated using the surface-water stage because the exchange perimeter extends from reach stage to the bottom of the layer or bottom of the reach (fig. 14).

Constant-Stage Reaches

Constant-stage reaches can be simulated by adding an additional flux term that compensates for inflows and outflows $(Q_M, Q_{PR}, Q_{EV}, Q_{LAT}, Q_{AQ}, Q_{BS})$ to a reach group. The constant stage boundary flux is

$$Q_{CS} = Q_{EV} - \left(\sum_{i\,1}^{nc} Q_M + Q_{PR} + Q_{LAT} + Q_{AQ} + Q_{BS} \right).$$

(52)

Constant-stage reaches ($\text{ISWRBND} < 0$) are not dynamic reaches and are solved explicitly after the continuity equation is solved for dynamic reaches. The interaction of dynamic reaches with constant-stage reaches is solved implicitly. The approach used to represent constant-stage reaches permits complete water budgets to be calculated for constant-stage reaches. The Q_{CS} term is zero for dynamic reaches simulated using the diffusive-wave approximation or the reservoir-routing approximation. Stage values for each constant-stage reach can vary by stress period or for each SWR1 time step if optional external time series data files are used.

Procedure for Specifying a Tilted Pool

With the reservoir-routing approximation, the water surface for reaches in a reach group can be tilted to approximate the water surface observed in the field. Although the tilt, or slope, of the water surface for each reach in a reach group is specified by the user, the vertical position of the water surface is determined using the simulated reach group stage solved as part of the solution to the continuity equation. The specified initial stages are used to calculate reach offsets, and thus the tilt of the water surface, for reach groups simulated using the reservoir-routing approximation. Calculated reach offsets are used to define the slope of non-equilibrated (tilted pool) reservoir-routing approximation reach groups. Reach offsets are not used for reach groups simulated using the diffusive-wave approximation. Reach offsets are defined relative to the reach with the lowest specified initial stage in the reach group, which is typically the farthest downstream. Reach offsets are calculated as

$$h_{offset_{ii}} = h_{initial_{ii}} - h_m,$$

(53)

where
   $h_{offset_{ii}}$ is the reach offset in reach ii,
   $h_{initial_{ii}}$ is the initial stage in reach ii (STAGE or STAGE2D), and
   h_m is the minimum initial stage for each reach in reach group m.

The approach used to calculate reach offsets is shown graphically in figure 15. As stated previously, only one stage value is calculated for each reach group; this value is calculated for the reach having the lowest initial stage within a given reach group. This approach is used only for reach groups composed of multiple reaches simulated using the tilted pool reservoir-routing approximation ($\text{IROUTETYPE} = 2$). The reach offset data are used to transform simulated stages from each reservoir-routing-approximation reach group using

$$h_{ii} = h_m + h_{offset_{ii}}.$$

(54)

Stages specified in subsequent stress periods are used to redefine reach offsets for dynamic reaches and stages for constant-stage reaches. Redefinition of reach offsets can be used to make seasonal changes in the slope of non-equilibrated reservoir-routing approximation reaches. For example, during the dry season, there may be no gradient in a reservoir, but there may be a gradient during the wet season when stormwater is being routed through the reservoir. Non-zero reach offsets would be specified using unequal stages in a reservoir-routing approximation reach. Specification of identical stages in a reservoir-routing approximation reach would result in a reach offset of zero and a flat water surface.

Water is conserved when reach offsets are modified in subsequent stress periods by adjusting stages at the beginning of the stress period based on reach volumes at the end of the previous stress period. This approach is valid as long as the stages in all adjusted tilted-pool reaches that had non-zero depths at the end of the last stress period have non-zero depths after adjustment.

Use of reach offsets can lead to conceptual difficulties and mass-conservation problems. For example, if initial stages are artificially low in the downstream reach and too high in an upstream reach, the reach offset in the upstream reach will be unrealistically large. As stages increase in reach group m, it is possible that stages in upstream reaches will exceed reasonable values as a result of erroneous reach offsets rather than reach-group inflows. Furthermore, if bidirectional structures are present in upstream reaches, it is possible that backflow will be incorrectly simulated. Information identifying the possibility of inaccurate flow at upstream structures in reservoir routing reach groups, reach-offset information, and volume errors associated with changes in reach offsets between subsequent stress periods are written to the MODFLOW listing (LST) file, but users are advised to carefully evaluate calculated reach offsets. The use of reach offsets is generally not recommended, and use of reaches simulated by the diffusive-wave approximation is preferred for streams, rivers, creeks and wetlands that do not equalize during a SWR1 time step.

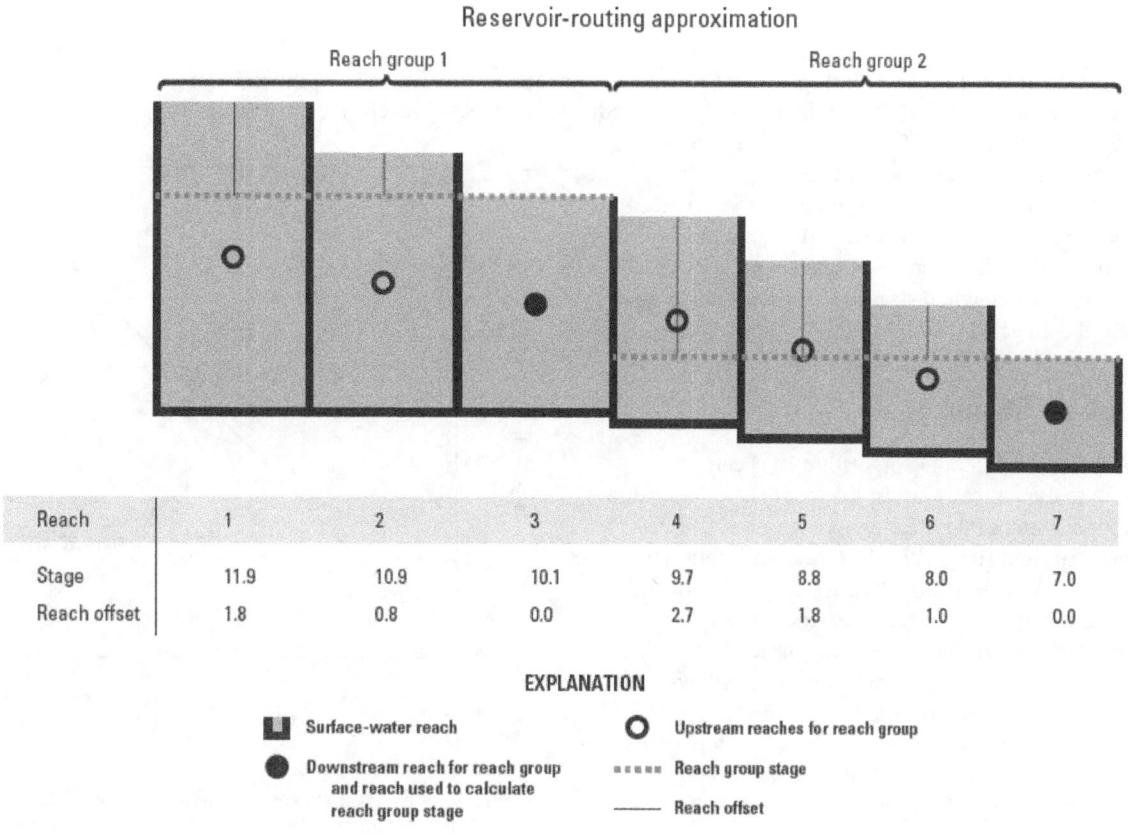

Figure 15. Approach used to calculate reach offsets for reservoir-routing reaches using the tilted-pool approach to simulate surface-water routing.

Numerical Controls

A number of options for controlling the SWR1 numerical solution have been included to allow users options for scaling flow terms for outflows from reach groups, and adaptive time stepping within the SWR1 Process. These options have been included to improve convergence of the solution and the quality of the Newton-step upgrade vector.

Scaling Reach-Group Outflows

The success of Newton schemes is dependent upon the presence of non-zero derivatives. Numerical instabilites can occur when reaches go dry because calculated derivatives (defined as the change in the reach flow residual divided by a stage perturbation) may equal zero. To avoid this problem, the SWR1 Process restricts reaches from going dry by maintaining a thin layer of water in each reach. This thin layer is maintained in the reach by scaling reach outflows so that the depth of water is never less than a user-defined minimum depth.

Q_M is scaled based on a user-defined minimum reach water depth (DMINDPTH), with units of length [L], using either a linear- (USE_LINEAR_DEPTH_SCALING option) or sigmoid-depth scaling function when the flow depth is greater than DMINDPTH, but less than or equal to an upper threshold depth, DUPDPTH. DUPDPTH is internally specified to be one order of magnitude larger than DMINDPTH. The linear-depth scaling function is

$$f_n = \frac{d - DMINDPTH}{DUPDPTH - DMINDPTH}.$$

(55)

The sigmoid-depth scaling function is

$$f_n = \frac{1}{1 + e^{\left(12\left(1 - \frac{d - DMINDPTH}{DUPDPTH - DMINDPTH}\right) - 6\right)}}.$$

(56)

The Q_M term in equation 18 is multiplied by f_n when d is less than or equal to DUPDPTH and greater than DMINDPTH. At water depths equal to DMINDPTH, f_n is zero and flow to adjacent connected reaches is not allowed ($Q_M = 0$). The linear- and sigmoid-depth scaling factors, f_n, are shown in figure 16.

To further minimize instabilities when depths are between DUPDPTH and DMINDPTH, reach evaporation and reach conductance are also scaled using the same linear- or sigmoid-depth scaling functions. Scaling is applied to reach conductance only when reach stages are greater than ground-water heads in connected model layers.

Adaptive Time Stepping

An adaptive time stepping algorithm has been implemented to dynamically adjust the length of SWR1 time steps. The algorithm has no effect on the lengths of MOD-FLOW time steps and stress periods. Use of this option will increase the temporal resolution during a MODFLOW time step when rainfall applied directly to a reach in a MOD-FLOW time step exceeds a user-defined value, when SWR1 convergence is not achieved, or when stages or inflows are changing quickly. The adaptive time stepping algorithm also includes provisions to increase the length of the SWR1 time step when reach rainfall is less than a user-specified value and stage and (or) inflow changes between subsequent SWR1 time steps are less than user-specified values. The maximum SWR1 time step, Δt_{RAI} for each MODFLOW stress period is defined using

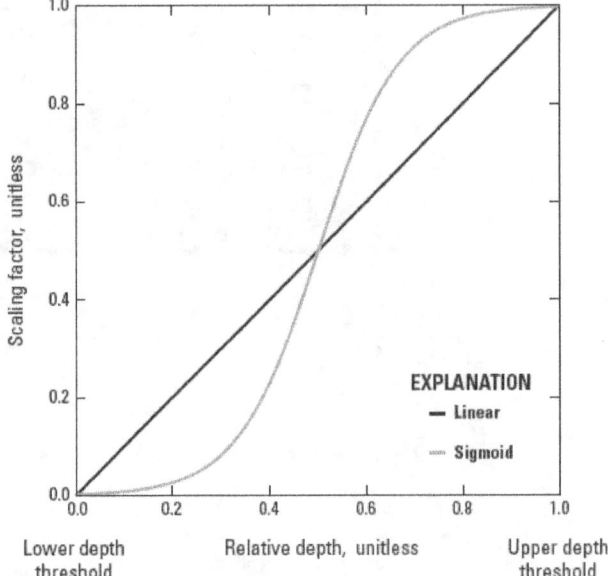

Figure 16. Scaling approach available in SWR1 to reduce instabilities related to diffusive-wave flow, evaporation losses, and aquifer-reach losses at small water depths. The lower depth threshold is assigned by the user and the upper depth threshold is specified to be one order of magnitude greater.

$$\Delta t_{RAI} = \max\left(\frac{\Delta t_{max}}{\dfrac{\max(\mathbf{RAI}_n)}{RAI_{max}}}, \Delta t_{min} \right),$$

(57)

where

Δt_{max} is the user-specified, maximum allowed SWR1 time-step length [T] (RTMAX),

$\mathbf{RAI}_n$ is the rainfall rate applied to SWR1 reaches in the current SWR1 time step,

RAI_{max} is the maximum rainfall rate allowed using Δt_{max} [L/T] (DMAXRAI), and

Δt_{min} is the user-specified, minimum allowed SWR1 time-step length [T] (RTMIN).

Equation 57 is used to preemptively reduce the length of the SWR1 time step, and it is used only if $\max(\mathbf{RAI}_n)$ is greater than RAI_{max}. If convergence was not achieved at the end of the current inner iteration, the SWR1 time step for the current outer iteration is reduced using

$$\Delta t^+ = \max\left(\frac{\Delta t}{\Delta t_{increment}}, \Delta t_{min} \right),$$

(58)

where

Δt^+ is the updated SWR1 time-step length for the current outer iteration [T],

Δt is the SWR1 time-step length used during the current outer iteration or previous SWR1 time step [T], and

$\Delta t_{increment}$ is a user-specified, SWR1 time-step adjustment factor [unitless] (RTMULT).

Stage changes are evaluated for each reach group using

$$f_h = \max\left(\frac{\left| \mathbf{h}^{t+\Delta t} - \mathbf{h}^t \right|}{\Delta h_{max}} \right),$$

(59)

where

f_h is the stage change factor [unitless],

$\mathbf{h}^{t+\Delta t}$ is the vector of all reach group stages [L] at the end of the current outer iteration,

$\mathbf{h}^t$ is the vector of all reach group stages [L] at end of the last SWR1 time step, and

Δh_{max} is the user-specified, maximum allowed stage change [L] between SWR1 time steps (DMAXSTG).

Inflow changes are evaluated for each reach group using

$$f_{Inf} = \max\left(\frac{\left| \mathbf{Inf}^{t+\Delta t} - \mathbf{Inf}^t \right|}{\Delta Inf_{max}} \right),$$

(60)

where

f_{Inf} is the inflow change factor [unitless],

$\mathbf{Inf}^{t+\Delta t}$ is the vector of all reach group inflows [L^3/T] at end of the current outer iteration,

$\mathbf{Inf}^t$ is the vector of all reach group inflows [L^3/T] at the end of the last SWR1 time step, and

ΔInf_{max} is the user-specified, maximum allowed inflow change [L^3/T] between SWR1 time steps (DMAXINF).

$\mathbf{Inf}$ in equation 60 is the sum of rainfall (Q_{PR}), positive reach inflows (Q_M), positive lateral flows (Q_{LAT}), positive aquifer-reach exchanges (Q_{AQ}), and positive boundary structure inflows (Q_{BS}) in equation 17. If $f_h > 1$ or $f_{Inf} > 1$ and convergence is achieved during the inner iterations of the current outer iteration, the SWR1 time step is reduced using

$$\Delta t^+ = \max\left(\frac{\Delta t}{\max(f_h, f_{Inf})}, \Delta t_{min} \right).$$

(61)

If SWR1 has converged for a user-defined number of SWR1 time steps (`NTMULT`), Δt for the current SWR1 time step is increased using

$$\Delta t^{t+\Delta t} = \min \left(\Delta t^t \Delta t_{\text{increment}}, \Delta t_{RAI}, \Delta t_{\max} \right),$$

(62)

where $\Delta t^{t+\Delta t}$ is the SWR1 time-step length for the current SWR1 time step. The adaptive time-step algorithm, if configured correctly, can decrease runtimes by minimizing the total number of SWR1 time steps needed during a simulation.

Aquifer Evapotranspiration Beneath Two-Dimensional Reaches

To simplify model construction and allow for evapotranspiration beneath model-based reaches (that is, reaches that cover the entire grid cell), a simple linear groundwater-evapotranspiration approach equivalent to the MODFLOW evapotranspiration (EVT) Package (Harbaugh, 2005) has been implemented. For cases in which the simulated reach stage is less than the upper threshold depth (`DUPDPTH`), reach evaporation will be less than the evaporation rate specified in the SWR1 input dataset. The simulated volumetric flow rate of reach evaporation, $Q_{EVactual}$, is calculated as the difference between the reach evaporation rate specified in the SWR1 input dataset and Q_{EV}. When $Q_{EVactual}$ is greater than zero, groundwater evapotranspiration, Q_{GWET}, is calculated as

$$Q_{GWET} = \begin{cases} (Q_{EV} - Q_{EVactual}) & \left(H_{j,i,k} > z_{ii} \right) \\ (Q_{EV} - Q_{EVactual}) \left(\dfrac{H_{j,i,k} - (z_{ii} - ETEXTD_{ii})}{ETEXTD_{ii}} \right) & \left((z_{ii} - ETEXTD_{ii}) \le H_{j,i,k} \le z_{ii} \right), \\ 0 & \left(H_{j,i,k} < (z_{ii} - ETEXTD_{ii}) \right) \end{cases}$$

(63)

where

z_{ii} is the bottom of reach ii [L], and
$ETEXTD_{ii}$ is the cutoff depth or extinction depth [L] (`GETEXTD`).

If evaporation is applied to model-based reaches, applied evaporation rates should be reference evapotranspiration rates characteristic of open water, and evapotranspiration should not be simulated with the MODFLOW EVT Package in these cells.

Procedure for Coupling the SWR1 Process with MODFLOW

Aquifer-reach exchanges and aquifer evapotranspiration beneath two-dimensional reaches are implicitly coupled at the time-step level within the MODFLOW outer (Picard) iterations. This approach of implicit coupling is identical to the coupling used with all other MODFLOW Packages, such as the SFR2, UZF1, and LAK Packages. Other approaches that could have been used for coupling the SWR1 Process with MODFLOW are (1) an explicit coupling, or (2) a fully implicit coupling with the groundwater and surface-water flow equations solved in the same matrix. For an explicit coupling, exchange terms would be lagged by one time step but this approach can lead to mass-conservation problems (Panday and Huyakorn, 2004) and under- and over-estimation of aquifer-reach exchanges. The fully implicit approach with groundwater and surface-water flow equations solved in the same matrix is preferred in many cases to ensure convergence and mass balance (Panday and Huyakorn, 2004), but this approach would require significant changes to the MODFLOW code structure. The approach used to couple aquifer-reach exchanges in the SWR1 Process avoids the problems introduced by explicit and fully implicit coupling, and thus represents a compromise between these two coupling approaches.

Because MODFLOW outer time-step iterations continue until convergence of the groundwater solution, groundwater heads and SWR1 flow terms are only out of sync by one outer iteration with one another. As a result, this coupling approach should result in convergence between the surface-water and groundwater system and mass conservation. To further enforce convergence between the surface-water and groundwater systems, an option for enforcing a user-defined aquifer-reach convergence criterion has been implemented. If aquifer-reach convergence is enforced in a MODFLOW simulation using the SWR1 Process, outer MODFLOW time-step iterations are continued until the maximum relative difference in SWR1 and MODFLOW aquifer-reach exchanges are less than a user-defined value, even if the groundwater solution has met its convergence criteria. Aquifer-reach convergence is defined as

$$\max\left|\frac{Q_{AQ_{ii}}^{k+1} - Q_{MF_{ii}}^{k+1}}{Q_{MF_{ii}}^{k+1}}\right| < f_{AQ_{max}}, \quad ii = 1, 2, ..., nr,$$

(64)

where

$Q_{AQ_{ii}}^{k+1}$ is simulated SWR1 aquifer-reach exchange for the current outer iteration for reach ii [L³/T],

$Q_{MF_{ii}}^{k+1}$ is simulated MODFLOW aquifer-reach exchange for the current outer iteration for reach ii [L³/T],

$f_{AQ_{max}}$ is the user-specified maximum allowed relative difference in SWR1 and MODFLOW aquifer-reach exchanges in any reach (TOLA), and

   nr is the number of reaches in the SWR1 dataset.

A generalized flowchart of the formulate component of the SWR1 Process within the groundwater flow (GWF) Process is shown in figure 17. Each subroutine is identified first by GWF2, followed by SWR1 (version 1 of the SWR Process), and then followed by the division of procedures used in MODFLOW. For example, GWF2SWR1AR is the module that allocates space for the SWR1 Process. The standard subroutines that are called from the main program are: GWF2SWR1AR, GWF2SWR1RP, GWF-2SWR1AD, GWF2SWR1FM, GWF2SWR1BD, and GWF2SWR1DA. The GWF2SWR1CV subroutine called from the main program determines if user-specified differences in aquifer-reach exchanges have been satisfied. A number of SWR1 Process specific

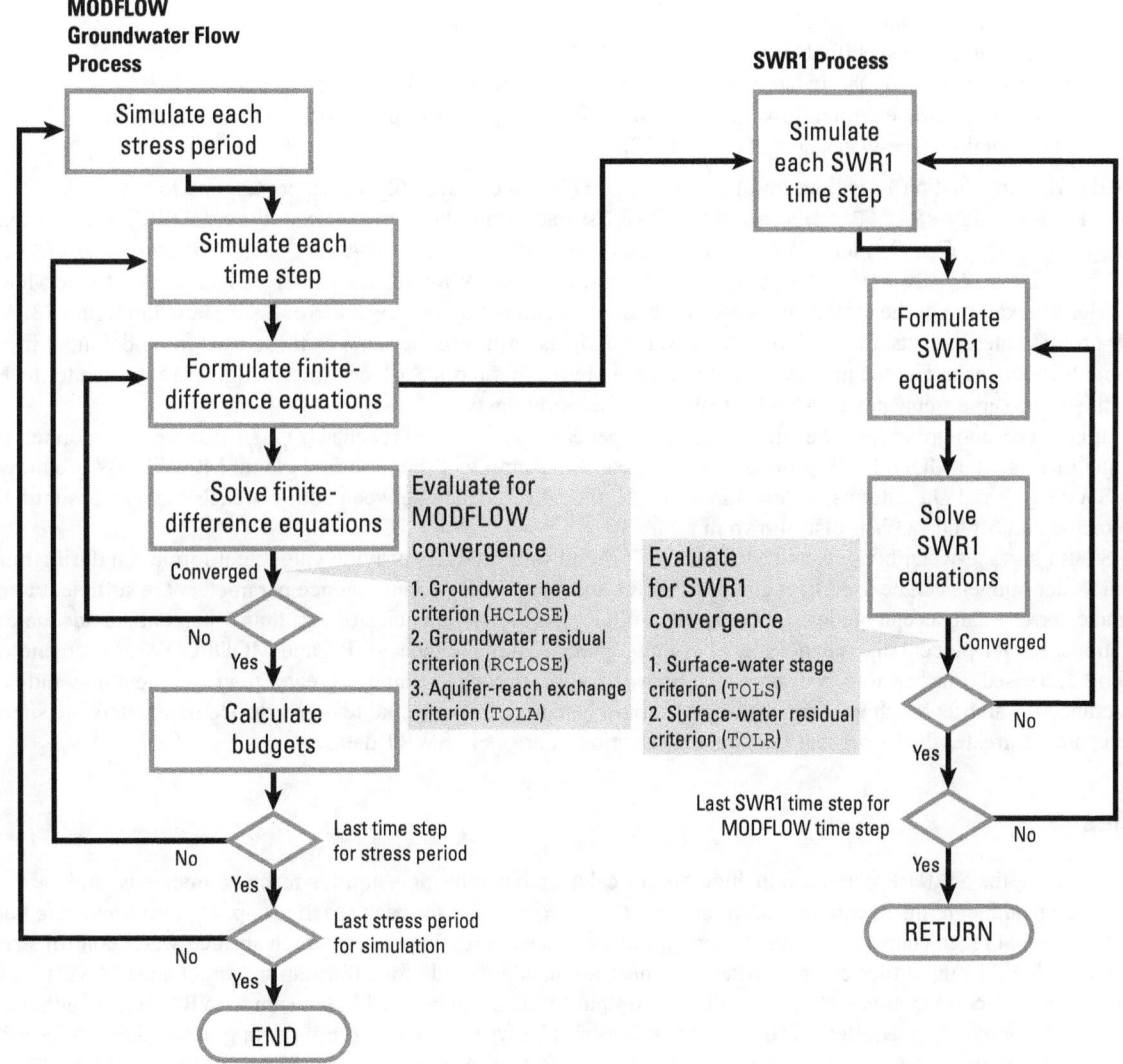

Figure 17. Generalized flowchart illustrating major components of the Surface-Water Routing (SWR1) Process in MODFLOW.

subroutines and functions are used internally by the SWR1 Process to process input data, compute flow terms, formulate the linear system of equations, and output results; SWR1-Process-specific subroutines and functions are identified first by S and followed by GWF2SWR or SWR. The SWR1 Process also calls a number of preconditioner and linear solver subroutines in the general linear solver library (gsol7.f) developed for the SWR1 Process.

Surface-Water Budgets

Global cumulative volumes [L^3] and incremental rates [L^3/T] of the surface-water budgets are calculated by evaluating the water budget on a reach-by-reach basis during each MODFLOW time step. Surface-water budget terms are calculated on a reach-by-reach basis to be consistent with the way MODFLOW groundwater budgets are calculated for individual MODFLOW Packages. Surface-water budget terms for each reach are segregated into positive and negative terms and accumulated into inflow (IN) and outflow (OUT) terms in each water budget. The SWR1 water-budget equation evaluated in each MODFLOW time step is

$$\frac{\Delta V}{\Delta t} + Q_{PR} + Q_{SL} + Q_{UZF} + Q_{EXT} - Q_{EV} + Q_{AQ} + Q_{BS} + Q_{CS} = Q_{error} \tag{65}$$

where

ΔV is the volume change [L^3],

Q_{SL} is the volumetric specified lateral flow rate [L^3/T],

Q_{UZF} is the volumetric dynamic lateral inflow (runoff) rate calculated by the UZF1 Package [L^3/T],

Q_{EXT} is the volumetric external flow rate programmatically applied using the GWF2SWR7EX subroutine [L^3/T], and

Q_{error} is the total surface-water budget error [L^3/T].

Q_{LAT} described earlier in equation 17 is the sum of Q_{SL}, Q_{UZF}, and Q_{EXT} in equation 65, which correspond to LATERAL FLOW, UNSAT. RUNOFF, and EXTERNAL FLOW terms in the SWR1 surface-water budget written to the MODFLOW LST file. SWR1 budget terms Q_{BS}, Q_{CS}, Q_{PR}, Q_{EV}, Q_{AQ}, and $\Delta V / \Delta t$ in equation 65 correspond to BOUNDARY FLOW, CONSTANT FLOW, RAIN-FALL, EVAPORATION, REACH-AQ FLOW, and STORAGE terms in the SWR1 surface-water budget written to the MOD-FLOW LST file. An example of the global surface-water budget calculated by the SWR1 Process is shown in figure 18. As with MODFLOW groundwater budgets, the units of SWR1 water-budgets terms are the same as those used in model input files (for example, cubic feet per second, cubic meters per second, or cubic feet per day). Surface-water budgets are written to the MOD-FLOW LST file at the same frequency as MODFLOW groundwater budgets.

Q_{AQ} in equation 65 and groundwater evapotranspiration beneath model-based reaches (Q_{GWET}) also are incorporated in the cumulative and incremental MODFLOW groundwater budgets. An example of the modified global MODFLOW groundwater budget with SWR1 Q_{AQ} and Q_{GWET} terms is shown in figure 19. The discrepancy between aquifer-reach exchanges simulated by the SWR1 Process and MODFLOW is also shown in figure 18.

Global SWR1 surface-water budgets and MODFLOW groundwater budgets provide valuable information during model development. Water budgets can be used to evaluate if SWR1 and MODFLOW convergence parameters are sufficient to reduce the mass-balance error to an acceptable level. In general, SWR1 percent discrepancies of less than 1 percent are adequate for most applications. SWR1 percent discrepancies greater that 1 percent may indicate SWR1 and MODFLOW convergence param-eters should be decreased. Furthermore, a discrepancy between aquifer-reach exchanges greater than 1 percent may indicate that a smaller tolerance for aquifer-reach exchanges (TOLA) should be defined. If reasonable convergence parameters are specified, percent discrepancies greater than 1 percent may be an indication of errors in SWR1 datasets.

Output Files

Output files from the SWR1 Process can include files of calculated reach stage, aquifer-reach components, and the summation of water-budget terms in equation 65 (including Q_M contributions) for each reach group. Q_M terms and calculated velocities for individual reach connections can also be output. Simulated structure data for each surface-water control structure can also be output. SWR1 output files can be written in American Standard Code for Information Interchange (ASCII) and binary output formats. A combination of ASCII and binary output formats can be used in the same SWR1 dataset, although data for one output type (for example, aquifer-reach components) cannot be written to both output data types in the same SWR1 simulation. ASCII data are written using a comma-separated-value format. A user-specified, SWR1 output print time (RTPRN) is used to control the frequency of ASCII and binary SWR1 output. Instantaneous SWR1 results at each output print time or the time-weighted average (SAVE_AVERAGE_RESULTS option) for each RTPRN time can be output. It is recommended that binary output formats be used for large SWR1 datasets.

```
                    VOLUMETRIC SURFACE WATER BUDGET FOR ENTIRE MODEL
                    AT END OF TIME STEP    1 IN STRESS PERIOD    2
------------------------------------------------------------------------------------

     CUMULATIVE VOLUMES      L**3        RATES FOR THIS TIME STEP      L**3/T
     ------------------                  ------------------------

            IN:                                     IN:
            ---                                     ---
       LATERAL FLOW  =        0.0000          LATERAL FLOW  =        0.0000
      UNSAT. RUNOFF  =      522.2346         UNSAT. RUNOFF  =      474.0138
           RAINFALL  =        0.0000              RAINFALL  =        0.0000
        EVAPORATION  =        0.0000           EVAPORATION  =        0.0000
       REACH-AQ FLOW =      474.2045          REACH-AQ FLOW =      400.8998
      EXTERNAL FLOW  =        0.0000         EXTERNAL FLOW  =        0.0000
      BOUNDARY FLOW  =        0.0000         BOUNDARY FLOW  =        0.0000
      CONSTANT FLOW  =     4168.4141         CONSTANT FLOW  =      150.9786
            STORAGE  =     7679.7573               STORAGE  =     4555.1953

          TOTAL IN  =    12844.6104             TOTAL IN  =     5581.0874

           OUT:                                    OUT:
           ----                                    ----
       LATERAL FLOW  =        0.0000          LATERAL FLOW  =        0.0000
      UNSAT. RUNOFF  =        0.0000         UNSAT. RUNOFF  =        0.0000
           RAINFALL  =        0.0000              RAINFALL  =        0.0000
        EVAPORATION  =     3483.8423           EVAPORATION  =     2592.1990
       REACH-AQ FLOW =     1909.7980          REACH-AQ FLOW =     1666.2837
      EXTERNAL FLOW  =        0.0000         EXTERNAL FLOW  =        0.0000
      BOUNDARY FLOW  =     6329.1465         BOUNDARY FLOW  =      191.3782
      CONSTANT FLOW  =      187.4492         CONSTANT FLOW  =     1124.6949
            STORAGE  =      922.2397               STORAGE  =        0.0000

         TOTAL OUT  =    12832.4756            TOTAL OUT  =     5574.5557

          IN - OUT  =       12.1348             IN - OUT  =        6.5317

  PERCENT DISCREPANCY  =        0.09    PERCENT DISCREPANCY  =        0.12


              DISCREPANCY BETWEEN MODFLOW AND SWR1 AQUIFER-REACH
                         TERMS FOR THE ENTIRE MODEL
------------------------------------------------------------------------------------
     CUMULATIVE VOLUMES      L**3        RATES FOR THIS TIME STEP      L**3/T
     ------------------                  ------------------------
               MODFLOW  =    1435.1067                  MODFLOW  =    1266.2734
                  SWR1  =    1435.5935                     SWR1  =    1265.3839
   PERCENT DISCREPANCY  =       0.03       PERCENT DISCREPANCY  =      -0.07
 SWR1 INACTIVE MODFLOW  =       0.0000    SWR1 INACTIVE MODFLOW  =       0.0000
------------------------------------------------------------------------------------
          PERCENT DISCREPANCY = 100 x (SWR1 - MODFLOW) / MODFLOW
```

Figure 18. Example cumulative and incremental surface-water budgets created by the SWR1 Process. The discrepancy between MODFLOW and SWR1 aquifer-reach exchanges is also shown.

SWR1 data written to each of the SWR1 output files include the following:

- **Time and stage data** (SSWR_P_RCHSTG subroutine)

  - Total time ($TOTIME$) [T]

  - SWR1 time-step length ($SWRDT$) [T]

  - MODFLOW stress period ($KPER$) [unitless]

  - MODFLOW time step ($KSTP$) [unitless]

  - SWR1 time step ($KSWR$) [unitless]

  - Reach stage ($STAGE$) [L]

```
          VOLUMETRIC BUDGET FOR ENTIRE MODEL AT END OF TIME STEP  1 IN STRESS PERIOD   2
          -----------------------------------------------------------------------------

              CUMULATIVE VOLUMES     L**3          RATES FOR THIS TIME STEP     L**3/T
              ------------------                   ------------------------

                    IN:                                     IN:
                    ---                                     ---
                   STORAGE =    1728.1625                  STORAGE =    3002.9216
             CONSTANT HEAD =       0.0000            CONSTANT HEAD =       0.0000
            HEAD DEP BOUNDS =      83.9826           HEAD DEP BOUNDS =     75.5151
              UZF RECHARGE =   25524.2812             UZF RECHARGE =     172.2364
                    GW ET =        0.0000                   GW ET =        0.0000
           SURFACE LEAKAGE =       0.0000          SURFACE LEAKAGE =       0.0000
               SWR LEAKAGE =    1909.5406              SWR LEAKAGE =    1666.5111
                  SWR GWET =       0.0000                 SWR GWET =       0.0000

                  TOTAL IN =   29245.9668                 TOTAL IN =    4917.1841

                   OUT:                                    OUT:
                   ----                                    ----
                   STORAGE =   23833.7109                  STORAGE =     136.4704
             CONSTANT HEAD =       0.0000            CONSTANT HEAD =       0.0000
            HEAD DEP BOUNDS =     183.8418           HEAD DEP BOUNDS =    106.5848
              UZF RECHARGE =       0.0000             UZF RECHARGE =       0.0000
                    GW ET =     4234.2134                   GW ET =     3836.3113
           SURFACE LEAKAGE =     503.6117          SURFACE LEAKAGE =     454.1677
               SWR LEAKAGE =     474.4340              SWR LEAKAGE =     400.2376
                  SWR GWET =       0.0000                 SWR GWET =       0.0000

                 TOTAL OUT =   29229.8105                TOTAL OUT =    4933.7715

                 IN - OUT =        16.1562                IN - OUT =      -16.5874

          PERCENT DISCREPANCY =      0.06    PERCENT DISCREPANCY =       -0.34
```

Figure 19. Example cumulative and incremental MODFLOW water budgets including SWR1 Process boundary fluxes with the groundwater model. SWR1 water-budget terms are highlighted.

- **Time and aquifer-reach data** (SSWR_P_QAQFLOW subroutine)

  - Total time (*TOTIME*) [T]

  - SWR1 time-step length (*SWRDT*) [T]

  - MODFLOW stress period (*KPER*) [unitless]

  - MODFLOW time step (*KSTP*) [unitless]

  - SWR1 time step (*KSWR*) [unitless]

  - Reach number (*REACH*) [unitless] – only written to ASCII output

  - Model layer (*LAYER*) [unitless]

  - Bottom elevation of the reach (*GBELEV*) [L]

  - Reach stage (*STAGE*) [L]

  - Reach depth (*DEPTH*) [L]

  - Groundwater head for model layer (*HEAD*) [L]

  - Reach wetted perimeter for model layer (*WETPERM*) [L]

  - Conductance for model layer (*CONDUCT*) [L^2/T]

- Calculated head difference for model layer (*HEADDIFF*) [L]

- Calculated aquifer-reach flow for model layer (*QAQFLOW*) [L³/T]

- **Time and reach group water-budget data** (SSWR_P_RGFLOW subroutine)

  - Total time (*TOTIME*) [T]

  - SWR1 time-step length (*SWRDT*) [T]

  - MODFLOW stress period (*KPER*) [unitless]

  - MODFLOW time step (*KSTP*) [unitless]

  - SWR1 time step (*KSWR*) [unitless]

  - Reach group (*RCHGRP*) [unitless] – only written to ASCII output

  - Reach group stage (*STAGE*) [L]

  - Net reach group inflow to connected reaches (*QPFLOW*) [L³/T]

  - Net reach group lateral flow (*QLATFLOW*) [L³/T]

  - Net reach group UZF1 inflow (*QUZFLOW*) [L³/T]

  - Net reach group rain (*RAIN*) [L³/T]

  - Net reach group evaporation (*EVAP*) [L³/T]

  - Net reach group aquifer-reach flow (*QAQFLOW*) [L³/T]

  - Net reach group outflow to connected reaches (*QNFLOW*) [L³/T]

  - Net reach group flow from external sources (*QEXTFLOW*) [L³/T]

  - Net reach group boundary structure flow (*QBCFLOW*) [L³/T]

  - Net reach group constant reach flow (*QCRFLOW*) [L³/T]

  - Net reach group volume change (*DV*) [L³/T]

  - Net reach group difference in inflows and outflows (*INF-OUT*) [L³/T]

  - Reach group volume (*VOLUME*) [L³]

- **Time and reach group connection lateral flow data** (SSWR_P_QMFLOW subroutine)

  - Total time (*TOTIME*) [T]

  - SWR1 time-step length (*SWRDT*) [T]

  - MODFLOW stress period (*KPER*) [unitless]

  - MODFLOW time step (*KSTP*) [unitless]

  - SWR1 time step (*KSWR*) [unitless]

  - Reach group (*RCHGRP*) [unitless] – only written to ASCII output

  - Reach number (*REACHC*) [unitless] – only written to ASCII output

  - Connected reach number (*CONNREACH*) [unitless] – only written to ASCII output

  - Net connection lateral flow (*QLATFLOW*) [L³/T]

- Net connection lateral velocity (*VLATFLOW*) [L/T] – based on the average water depth at the connection for connections with and without a control structure

- **Time and structure flow data** (`SSWR_P_STRFLOW` subroutine)

    - Total time (*TOTIME*) [T]

    - SWR1 time-step length (*SWRDT*) [T]

    - MODFLOW stress period (*KPER*) [unitless]

    - MODFLOW time step (*KSTP*) [unitless]

    - SWR1 time step (*KSWR*) [unitless]

    - Reach number (*REACH*) [unitless] – only written to ASCII output

    - Structure number (*RCHGRP*) [unitless] – only written to ASCII output

    - Upstream reach stage (*USSTAGE*) [L]

    - Downstream reach stage (*DSSTAGE*) [L]

    - Gate elevation (*GATEELEV*) [L]

    - Gate opening (*OPENING*) [L]

    - Structure flow (*STRFLOW*) [L^3/T]

Q_{AQ} and Q_{GWET} data are written in standard three-dimensional binary formats when cell-by-cell fluxes are requested. The storing frequency of SWR1 data to cell-by-cell data files is the same as all other MODFLOW cell-by-cell flux data, and is specified in the MODFLOW Output Control (OC) data file. The unit number of the cell-by-cell file to use for SWR1 data is specified in the SWR1 data file in a manner consistent with other MODFLOW Packages.

Cell-by-cell output from the SWR1 Process can be post-processed using available post-processors and tools, because standard MODFLOW functionality is used to create these output data. Comma-separated-value ASCII output files can be easily post-processed in standard spreadsheet programs. Standard tools are not available to post-process binary SWR1 surface-water data output files. However, the data format used for binary SWR1 data-output files is simple, and will facilitate development of post-processing tools to work with binary SWR1 output data. Stage data, aquifer-reach data, reach group water budget data, and reach group connection lateral flow data are written using the `SSWR_P_RCHSTG`, `SSWR_P_AFLOW`, `SSWR_P_RGFLOW`, and `SSWR_P_QMFLOW` subroutines, respectively; readers are referred to these subroutines for additional information on SWR1 data output formats.

Routing Option Assumptions and Limitations

The SWR1 Process can simulate surface-water flow in a network of interconnected one-dimensional channels, two-dimensional features, and generic one- and two-dimensional features with unsteady, bidirectional flow. The diffusive-wave approximation is one option for routing surface water between reach groups and is based on a simplified version of the Saint-Venant equations; therefore, it should be used appropriately. The diffusive-wave approximation of the Saint-Venant equations is a good approximation for systems in which flood-wave dissipation by inertial forces or downstream control is negligible (Ponce, 1991). This approximation is valid in systems where the Froude (*Fr*) number is low (Ponce, 1990). The Froude number for a rectangular channel is

$$Fr = \frac{v}{c} = \frac{v}{\sqrt{gd}} \sim \frac{\text{inertial force}}{\text{gravitational force}}, \tag{66}$$

where

$\quad v \quad$ is the flow velocity [L/T], and

$\quad c \quad$ is the wave celerity [L/T] for cases where the change in water depth is small relative to the water depth.

The Froude number is a measure of the ratio of the flow velocity and the wave celerity, also defined as the wave velocity (Chaudhry, 2008). Subcritical, critical, and supercritical flow occurs at Froude numbers <1, 1, and >1, respectively.

A disturbance can propagate upstream under subcritical flow conditions; this condition is referred to as downstream controlled (Henderson, 1966). Under critical or supercritical flow conditions flow cannot be influenced by a downstream disturbance; this condition is typically referred to as upstream controlled (Henderson, 1966). The diffusive-wave approximation always results in upstream propagation of a disturbance when an upstream stage gradient exists, regardless of the flow conditions. As a result, the diffusive-wave approximation is applicable to subcritical flow conditions (that is, the Froude number is less than 1).

Ponce (1990) determined that at low Froude numbers, the diffusive-wave approximation defined in equation 9 is a good approximation of a generalized diffusive-wave approximation that includes convective and local inertial effects. The generalized diffusive-wave approximation developed by Ponce (1990) that includes inertial effects is

$$\frac{\partial d}{\partial t} + \frac{3}{2}v_o\frac{\partial d}{\partial x} = \left(1 - \alpha Fr^2\right)D_h\frac{\partial^2 d}{\partial x^2} = \tilde{D}_h\frac{\partial^2 d}{\partial x^2},$$

(67)

where

v_o is a reference mean flow velocity [L/T],
α is a factor that defines the relation of the Froude number and inertial terms [unitless],
D_h is the noninertial hydraulic diffusivity [L²/T], and
$\tilde{D}_h$ is the actual (inertial) hydraulic diffusivity [L²/T].

The term α is equal to zero for the noninertial diffusive-wave approximation, -0.5 for the diffusive-wave approximation including convective inertia, 0.75 for the diffusive-wave approximation including local inertia, and 0.25 for the full inertial diffusive-wave approximation (Ponce, 1990). The noninertial hydraulic diffusivity based on reference flow (Hayami, 1951) is defined as

$$D_h = \frac{v_o d_o}{2S_o},$$

(68)

where

d_o is a reference flow depth [L], and
S_o is the bottom slope [L/L].

Errors introduced by ignoring inertial terms are Froude-number dependent and, in general, increase wave dissipation and propagation. As a result, simulated stages are decreased at a disturbance (wave crest) and increased away from a disturbance when the diffusive-wave approximation is used and inertial terms are large. For example, using equation 67 and $\alpha = 0.25$ would result in an actual (inertial) hydraulic diffusivity, $\tilde{D}_h$, that is a factor of 0.75, 0.86, and 0.98 smaller than D_h at a Froude number of 1.0, 0.75, and 0.25, respectively.

Comparison of simulated water depth for several problems using the dynamic- and diffusive-wave approximations of the Saint-Venant equations are shown in figure 20A. These problems start with an initial water-level elevation of 2.05 m, and have the same horizontal grid-cell dimensions and bottom elevations. The only differences between the simulations are the inflow rate and Manning's roughness coefficients calculated using Manning's equation (eq. 6) to result in a steady-state water depth of 1.0 m. Although all of the simulations result in the same steady-state water depth, there are significant discrepancies between results simulated using the dynamic- and diffusive-wave approximations at simulation times of less than 1 hour for models using Manning's roughness coefficients less than 0.10 s/m^(1/3) (fig. 20B). Calculated maximum changes in the Froude number with time using the diffusive-wave approximation and the resultant hydraulic diffusivity correction factor, $1 - \frac{1}{4}Fr^2$, of Ponce (1990) for the Manning's roughness coefficients evaluated are summarized in table 4. Results for this particular problem indicate stage differences in excess of 1 percent occur at Froude numbers greater than 0.15 and hydraulic diffusivity correction factors less than 0.99. Based on equation 67 with an $\alpha = \frac{1}{4}$, D_h and $\tilde{D}_h$ will differ 1 percent or less for Froude numbers less than or equal to 0.2; this result is consistent with the results shown in figure 20A and summarized in table 4.

The Manning's equation (eq. 6) and Froude number equation (eq. 66) can be used to solve for the water depth (d) at a specified Froude number. Figure 20B shows the relation of the water surface depth to the water-surface slope for a Froude number of 0.20 using Manning roughness coefficients ranging from 0.001 to 0.500 s/m^(1/3). The slope of the lines in figure 20B have a constant value of -3 because of the exponents of the water depth in the Manning's equation (eq. 6), the water depth in the wave celerity equation (eq. 66), and the water-surface elevation in the Manning's equation (eq. 6). The horizontal offset of the lines in figure 20B is a function of $\log\left(Fr \cdot n \cdot \sqrt{g}\right)$.

The combination of the relation of the water-surface depth to water-surface slope (fig. 20B) and the results shown in figure 20B can be used to estimate conditions when use of the diffusive-wave approximation would differ significantly from the dynamic-wave approximation. For point a in figure 20B, it is expected that diffusive-wave approximation would result in errors

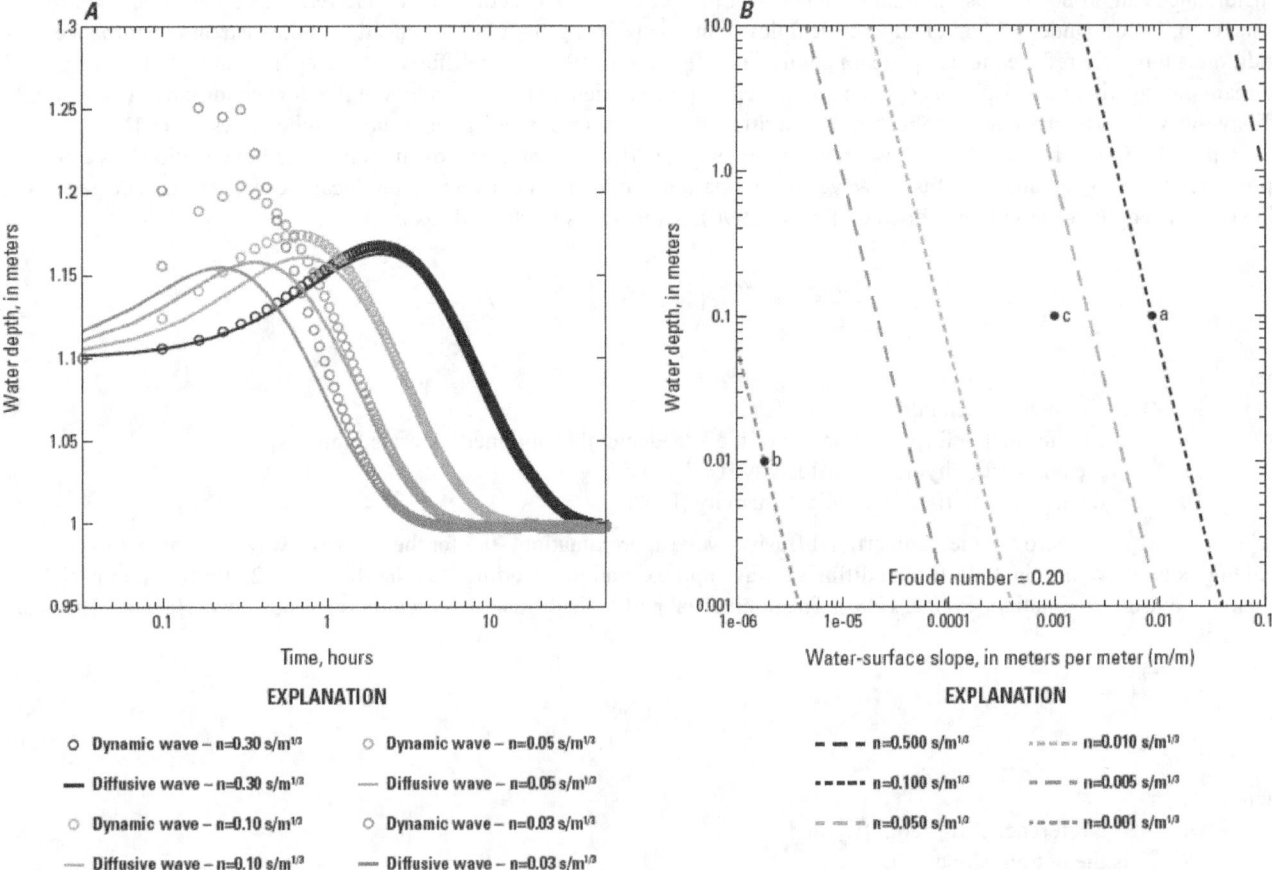

Figure 20. *A*, Comparison of simulated water depth for four simulations with varying steady-inflow conditions and Manning's roughness coefficients (*n*) using dynamic- and diffusive-wave approximations of the Saint-Venant equations, and *B*, the relation between the water-surface slope and water depth for a variety of Manning's roughness coefficients at a Froude number of 0.2. Points a, b, and c in *B*, are explained in the text.

Table 4. Calculated SWR1 maximum difference in Froude numbers with time and associated hydraulic diffusivity correction factors calculated using the diffusive-wave approximation.

Manning's roughness coefficient, in seconds per meter $^{1/3}$	Simulated maximum Froude number difference with time	Hydraulic diffusivity correction factor, unitless
0.30	0.049	0.999
.10	.149	.994
.05	.275	.981
.03	.441	.951

less than 1 percent ($Fr = 0.2$) at a water depth of 0.1 m, water-surface slopes less than 8.5×10^{-3} m/m, and Manning roughness coefficients greater than or equal to 0.1 s/m$^{1/3}$. For point b in figure 20*B*, diffusive-wave approximation errors would be less than 1 percent ($Fr = 0.2$) for a water depth of 0.01 m, water-surface slopes less than 1.8×10^{-6} m/m, and Manning's roughness coefficients greater than or equal to 0.001 s/m$^{1/3}$. Diffusive-wave approximation errors would be less than 1 percent ($Fr = 0.2$) for a water depth of 0.1 m and water-surface slopes less than 1.0×10^{-3} m/m (point c) if the Manning's roughness coefficient was 0.034 s/m$^{1/3}$ or greater. Because of the inverse relation between the product of the water-surface slope and depth to the Froude number (eq. 66), smaller Froude numbers correspond to larger Manning's roughness coefficients for a given water-surface slope and depth. For example, at a Froude number of 0.02, the line intersecting point a in figure 20B would have a Manning's roughness coefficient of 1.0 s/m$^{1/3}$.

Figure 20B can be used prior to a simulation by estimating maximum water-surface slopes and associated water depths in uncontrolled reaches. When these points are plotted in figure 20B, they can be used to determine if the diffusive-wave approximation is expected to have an error of 1 percent or less by evaluating their position relative to the specified Manning's roughness coefficient for the reach. For example, errors greater than 1 percent might occur if the Manning's roughness coefficient of the plotted reach were less than or equal to the value of the nearest line to the left of the point (for example, point c in figure 20B) if the reach Manning's roughness coefficient was 0.01 s/m$^{1/3}$.

After a simulation is run, maximum Froude numbers should be evaluated (ISWRPFRN > 0) to confirm that they are less than or equal to 0.2. Simulations with Froude numbers greater than 0.2 should be further evaluated to justify use of the diffusive-wave approximation for this surface-water system. Although there may be significant differences between the diffusive- and dynamic-wave approximations for a given surface-water system, it may still be reasonable to use the diffusive-wave approximation if the periods where the Froude number exceeds 0.2 do not exceed a significant percentage of the total simulation time.

In some cases, use of the SWR1 Process may not be appropriate for a given application because the assumptions used to develop the diffusive-wave approximation are violated (acceleration terms are not negligible). For these applications, the dynamic-wave approximation should be used. Use of the dynamic-wave approximation could be accomplished by modifying the SWR1 Process to solve for both continuity and momentum or coupling a code that solves the dynamic-wave approximation. An example would be the coupling of the BRANCH code (Schaffranek and others, 1981) and MODFLOW in MODBRANCH (Swain and Wexler, 1996).

Although it is more difficult to estimate errors for the reservoir-routing approximation, it is expected that stage errors could exceed the errors of the diffusive-wave approximation routing option because all inflows are equilibrated over the reach group. Similar to the diffusive-wave approximation, it is expected that stage errors would be greatest during inflow events and decrease thereafter. The diffusive-wave approximation would be the preferred approach to use in reaches where it is not reasonable to equilibrate inflows over all of the reaches in a reach group.

Aquifer-Reach Exchange Assumptions and Limitations

The assumption that aquifer-reach exchanges are transmitted to the water table instantaneously when the groundwater head is below the bottom elevation of the reach generally limits the applicability of the SWR1 Process to the simulation of transient leakage through thin unsaturated or nearly saturated zones beneath reaches. This assumption may not be reasonable for surface-water systems in which the water table is tens to hundreds of feet or meters below the bottom of the reach. In addition, when the water table is below the bottom of the reach, aquifer-reach exchange is dependent upon the difference between the stage in the reach and the reach bottom, and the rate of leakage is not dependent on the vertical hydraulic properties of the underlying aquifer.

The restriction that each reach can interact only with groundwater-model cells in the same row and column results in some limitations on cell size compared to the dimensions of the reach. First, the maximum width of a reach should be no greater than that of a finite-difference cell. The model will still operate if this assumption is violated, but some accuracy will be lost because the groundwater seepage will occur in a single cell when it should be divided among all the cells covered by the width of the reach, and non-convergence may occur with large conductance values. Similarly, in the vertical direction, the reach geometry should not cross through more than one model layer unless the option for automatically connecting reaches to multiple layers is used. Again, the model will operate if this assumption is violated, but the groundwater seepage will be represented by only a single layer.

The appropriateness of reaches that fully penetrate one or more layers should be carefully considered because these reaches represent a physical boundary for the groundwater system, that is, the physical edge of the aquifer. This physical separation of the aquifer on either side of these reaches will not be represented correctly in the model. This limitation would also apply to MODFLOW river (RIV) boundary conditions representing reaches fully-penetrating one or more layers, and is a consequence of the boundary condition discretization scheme used in MODFLOW.

Tips for Designing SWR1 Models

Getting started with any new MODFLOW Process or Package requires an investment of time and effort to become familiar with the intricacies of that particular process or package. Although the SWR1 Process was designed to be relatively easy to use, new users may encounter some problems in getting the program to work properly. The purpose of this section is to discuss some of the potential problems that users may face when designing a SWR1 dataset, and provide tips for avoiding the most common errors.

Perhaps the most difficult part of developing a SWR1 dataset is creating the discretized reaches and specifying reach connectivity (NCONN and ICONN). In many cases, the SWR1 Process will run even if the connectivity is incorrect. However, without specification of the correct connectivity, simulated flows and stages will be incorrect. Use of the SWRPre preprocessing program (described in app. 4) can automate development of discretized reaches with the correct connectivity from available Esri polyline shapefiles and MODFLOW discretization files. Reach connectivity and discretization also are important when control structures are simulated. It is good practice to test the surface-water component without the groundwater component to ensure that the model runs in surface-water-only mode prior to running coupled surface-water/groundwater simulations. Surface-water-only simulations can be run by setting the ISWRONLY variable in a SWR1 dataset to a value greater than 0; this setting internally resets the boundary (IBOUND) array in the MODFLOW Basic (BAS) Package to 0, which disables the groundwater flow (GWF) Process and SWR1 aquifer-reach exchanges.

Perhaps the best approach for designing a new SWR1 dataset is to start with a simple numerical representation of the surface-water network that uses constant-stage reaches (ISWRBND < 0 for all active reaches) and runs relatively quickly. By designing a model in this manner, the user creates a simple tool that can be used to quickly evaluate appropriate grid resolution, aquifer-reach exchange parameters, and computer runtimes. The constant-stage approach also can also be used for surface-water systems with control structures. SWRPre also can be used to simplify evaluation of different horizontal grid resolutions. The simple SWR1 dataset can then be used throughout the development of larger, more detailed model datasets with increasing complexity. Development of a simple SWR1 dataset with constant-stage reaches is strongly recommended for first-time users of the SWR1 Process.

When considering increasing the complexity of the SWR1 dataset and using the diffusive-wave approximation (IROUTETYPE = 3) to simulate surface-water flow, users are advised to estimate Froude numbers prior to performing a simulation using observed water depths, observed water-surface slopes, and channel properties. An analysis similar to the analysis shown in figure 20B can be used to estimate Froude numbers from observation and channel geometry data and estimate if stage errors resulting from ignoring inertial terms are expected to be significant. It is expected that surface-water stage errors will be less than 1 percent for Froude number less than or equal to 0.2. SWR1-calculated Froude numbers (ISWRPFRN > 0) should be evaluated after each model simulation to confirm that the amount of time Froude numbers exceed 0.2 represents a small proportion of the total simulation time. The reservoir-routing approximation in the SWR1 Process can also be used, but may have errors that equal or exceed errors from the diffusive-wave approximation as a result of equilibrating inflows over the entire reach group. Another approach (for example, a Process or Package that implements the dynamic-wave approximation) should be used if the diffusive-wave or reservoir-routing approximation options are expected to have a significant effect on simulated stages and aquifer-reach exchanges.

One important consideration in the design of a dynamic surface-water routing process is the selection of appropriate numerical grid resolution. Reach groups with small lengths generally require small SWR1 time steps to accurately simulate surface-water flow processes. Combining small reaches that are connected to larger reaches, or multiple connected small reaches, into a single reach group is an approach that can minimize the SWR1 time step requirements. At present, there is no way to determine the required level of resolution in a SWR1 dataset prior to doing an actual simulation. Experience has shown that SWR1 networks designed with approximately uniform reach-group lengths are less prone to numerical instabilities than models designed with widely varying lengths. SWRPre can be used to develop SWR1 networks with different reach-group lengths to facilitate determination of reach group lengths that minimize runtimes and do not adversely affect simulated surface-water results.

Selection of appropriate time-step lengths for integrated surface-water/groundwater models can be challenging because surface-water and groundwater processes respond at different timescales. Smaller surface-water time steps are generally necessary during and immediately after high-intensity events to accurately simulate floodwave propagation. Larger surface-water time steps are possible during times characterized by low-intensity lateral inflows or distributed inflows, such as aquifer-reach exchanges. The ability to have multiple SWR1 time steps in each MODFLOW time step can reduce the difficulty of selecting a single appropriate time-step length. SWR1 time steps ranging from 1 minute to 1 hour are typical time-step lengths during high-intensity events. It is recommended that constant MODFLOW time-step lengths be used when dynamic SWR1 reaches are simulated. Daily MODFLOW time steps with 1-minute SWR1 time steps are common in SWR1 simulations during high-intensity events.

As previously described, a model-based reach is one that covers an entire MODFLOW cell. Model-based reaches allow surface-water features such as lakes or wetlands to be represented, but care must be taken to ensure that boundary conditions associated with these reaches are consistent with other MODFLOW Packages. If rainfall is applied to model-based reaches in the SWR1 dataset, then it should not be applied in the MODFLOW Recharge (RCH) Package to these groundwater cells. If evaporation is applied to model-based reaches, then applied evaporation rates should be reference evapotranspiration rates characteristic of open water, and evapotranspiration should not be simulated with the MODFLOW EVT Package in these groundwater cells. With the SWR1 Process, groundwater evapotranspiration (Q_{GWET}) from underlying model layers will be calculated for cells beneath model-based reaches when the evaporation from the surface-water system is less than reference evaporation rates.

Because rainfall and evapotranspiration will affect the surface-water system before affecting the groundwater system, it is recommended that rainfall and evapotranspiration be simulated using the SWR1 Process for model-based reaches instead of applying the MODFLOW RCH and EVT Packages to underlying aquifer cells.

The concept of steady-state or transient conditions can be confusing when referring to surface-water models because of the time required for surface-water volumes to equilibrate with static groundwater levels and steady-state boundary conditions (for example, surface-water inflow, rainfall, and evapotranspiration). Surface-water flow is considered to be steady state when stage values do not change with time.

To achieve steady-state conditions, MODFLOW users commonly invoke the steady-state flag for one or more stress periods as an alternative to running long transient stress periods. It is possible to do this with MODFLOW and the SWR1 Process, but it is numerically challenging because of the nonlinear stage-volume relation characteristic of most surface-water features; groundwater cells typically have a linear groundwater head-volume relation. There are two options for simulating steady-state conditions in a single SWR1 stress period. The first, and default, option is to neglect storage changes in the surface-water system as is done in MODFLOW with the steady-state flag. The second option is to use a pseudo-transient continuation approach, which considers storage changes in the surface-water system during each nonlinear and linear iteration, but continues each SWR1 time step until storage changes are less than the flow convergence criterion (TOLR). The pseudo-transient continuation option is used if the USE_STEADYSTATE_STORAGE option is specified. For both options, a stable single-time step steady-state solution will require either dampening or backtracking (but not both). The aquifer-reach convergence criteria (TOLA) should be set to a small number (for example, 1×10^{-6} m^3/s), and the number of MODFLOW outer iterations should be set to a large number (for example, 1,000) for single-time step steady-state simulation with SWR1. For the cases evaluated, neglecting storage changes (the default option) and dampening ($0.0 < DAMPSS < 0.5$) have performed well. The single-time-step steady-state simulation may work best if started with relatively high initial stages rather than low-to-dry initial stage conditions. In general, the single-time-step steady-state solution that does not consider storage change executes SWR1 time steps faster than the pseudo-transient continuation approach.

Another approach for achieving steady-state surface-water conditions is to do a transient model run for a long simulation period (for example, 10 years) with constant hydrologic stresses until stages and groundwater levels do not change with time. At this point, the model has come to steady state with respect to surface-water and groundwater flow. Users are advised to confirm that stages converge to approximately the same values from different initial conditions to ensure that steady-state surface-water conditions have been achieved. Although there are a number of ways to determine when the surface-water model has reached steady state conditions, perhaps the easiest way is to plot simulated stages or flows in the model as a function of time.

In many coupled surface-water/groundwater simulations, the goal is to evaluate temporal changes in stage, surface-water flow, groundwater head, and groundwater flow. For most of these types of simulations, it is important that the initial stages and groundwater heads are in equilibrium with one another and that they have reached consistency with the imposed hydrologic stresses. If initial stages and groundwater heads are not in equilibrium, simulated transient changes may be the result of equilibration of stages and groundwater heads to external boundary conditions rather than to changes in internal hydrologic stresses (rainfall, evaporation, and lateral flows). Two types of simulations can be used to produce appropriate initial conditions for a transient model. The first is a simulation that has been run to steady state using representative hydrologic conditions from that time period or a hydrologically similar time period. The other is a transient simulation with temporally varying stresses that has been run repeatedly, each time with the initial conditions specified based on the results of the previous run, until the model produces the same results each time. The user should determine the best approach for the particular problem.

There are a number of options (ISOLVER) available for solving the surface-water flow equation. Selection of the best method depends primarily on the particular problem and usually involves a compromise between solution accuracy and length of computer runtime. For example, the direct solution methods generally result in the best solution, but this approach can have high numerical overhead. As a result, the direct solvers should only be applied to surface-water networks having fewer than 1,000 reach groups. The iterative solvers work well for large surface-water networks but are more sensitive to convergence criteria. It is possible to use exact or inexact Newton methods with the iterative solvers. The inexact Newton method (USE_INEXACT_NEWTON option) is preferred because convergence is generally faster than the default exact Newton method. However, there may be some SWR1 problems that benefit from strict enforcement of TOLR during every call to the iterative solvers. Appendix 3 provides details on the exact and inexact Newton methods. A number of preconditioners (IPC) are available for use with the iterative solvers. In some cases, preconditioners can improve convergence and reduce runtimes. Users are advised to experiment with all of the methods and options to find the one that provides reasonable results with the shortest runtimes. Extensive experimentation with the available solvers indicates that the biconjugate gradient stabilized (ISOLVER=2) with the MILU(0) preconditioner (IPC=3) with a stage (TOLS) and flow (TOLR) convergence criteria of 1×10^{-9} m and 0.01 m^3/s, respectively, works well for most simulations. The MILU(0) preconditioner is a modified incomplete factorization for non-symmetric matrices (Trefethen and Bau, 1997; Saad, 2003) and is comparable to the modified incomplete Cholesky (MIC) preconditioner used in the Preconditioned Conjugate Gradient (PCG) Package for MODFLOW (Hill, 1990). It has been determined through experimentation that application of MIC and MILU(0) to symmetric matrices results in preconditioned matrices of comparable quality and convergence properties. The Jacobi (IPC=1) or ILU(0) (IPC=2) preconditioners should be used in cases where the MILU(0) preconditioner does not work well. An incomplete level fill preconditioner (IPC=4) with a dual drop strategy (ILUT) based on Saad (1994a) from the SPARSKIT library (Saad, 1994b) has also been included for extremely difficult problems. The ILUT preconditioner is computationally expensive and should only be used in cases where the other preconditioners do not work well.

A line-search algorithm (IBT > 1) has been included with both the direct and iterative solvers and can improve convergence and the quality of the Newton-step upgrade vector for highly nonlinear surface-water problems. There is a computational cost associated with the line-search algorithm, so it should only be implemented if the problem is difficult to solve without line search and (or) line search noticeably reduces runtimes. The adaptive time-stepping algorithm (RTMIN < RTMAX and RTMULT > 1.0) can be used to reduce runtimes by allowing stage and flow conditions to control SWR1 time steps. With daily MODFLOW time steps, the adaptive time-stepping algorithm could be set with a minimum and maximum SWR1 time-step length of 1 minute and 1 day, respectively. It is possible to set the adaptive time-stepping rainfall (DMAXRAI), stage (DMAXSTG), and inflow (DMAXINF) control parameters too small, which can result in longer runtimes than might result from use of a constant SWR1 time-step length of 1 minute. As a result, adaptive time stepping should only be used if the smallest time-step length needed during a simulation is several orders of magnitude less than the maximum permitted time-step length (for example, 1 minute versus 100 minutes). Adaptive time-stepping stage and inflow control parameters should be adjusted for the particular SWR1 dataset.

One of the most time-consuming problems that some users may encounter with SWR1 is trying to achieve convergence, which requires achieving an acceptable surface-water mass balance (TOLR) and possibly satisfying aquifer-reach convergence criteria (TOLA). In many instances, convergence problems are due to errors in the input files, but once found, the errors can be easily corrected. Some of the most common errors in the input files are often the result of

1.   Using inconsistent units for surface-water flows,

2.   Using a different vertical datum for the surface-water and groundwater components,

3.   Using incorrect spatial (DLENCONV) and temporal (TIMECONV) conversion parameters in SWR1 datasets with control structures, and (or)

4.   Specifying SWR1 time steps that are too large for high-intensity events.

Complex models and conceptual errors in model design can also cause problems with convergence. For example, problems with wetting and drying will often cause convergence problems for surface-water/groundwater coupling terms, as will some water-table problems and conditions with layer conversions from confined to unconfined. Some common conceptual errors that may cause convergence problems include using inappropriate initial conditions, applying rapidly changing boundary conditions with insufficient temporal discretization, or assigning drastically different values to reach and aquifer properties to adjacent reaches and groundwater-model cells.

The ability to continue a simulation when convergence is not achieved in the MODFLOW groundwater-flow process and (or) the SWR1 Process has been included as an option (non-convergence continuation option) and is implemented by specifying the USE_NONCONVERGENCE_CONTINUE option. Use of the non-convergence continuation option can be beneficial for troubleshooting SWR1 datasets and determining appropriate convergence criteria for specific SWR1 datasets. MODFLOW time steps that did not meet specified convergence criteria are reported to the MODFLOW listing file. Users should evaluate surface-water and groundwater mass balance for MODFLOW time steps that did not meet specified convergence criteria if the non-convergence continuation option is used.

SWR1 results that are typically evaluated at the end of a run include the ASCII or binary SWR1 output files that contain stages, reach group and aquifer-reach flow terms, and cell-by-cell flows. Users are advised to verify that mass-balance errors summarized in the MODFLOW listing (LST) file are within an acceptable range. Users also should review time series plots of stages and surface-water flows at critical locations. When a model has been designed properly and SWR1 produces an accurate solution, users should expect to see stages that decrease in the downstream direction. To facilitate a more accurate evaluation of simulated Froude numbers, a summary of the location of the maximum Froude number at the end of each time step can be output to the LST file (ISWRFRN > 0). When first running a SWR1 dataset with reaches simulated by the diffusive-wave approximation, the Froude number output should be enabled and evaluated to confirm that estimated maximum Froude numbers are comparable to maximum simulated Froude numbers and errors induced by inertial terms are within a user-defined range.

Test Simulations

Five hypothetical simulations are used to illustrate the capabilities of the SWR1 Process. The first four test simulations are surface-water-only simulations that are compared to existing solutions or other surface-water simulators. The last test was developed to test aquifer-reach interaction and different surface-water geometry types, reach conductance options, control structure types, and SWR1 options for simulating surface-water routing. Detailed results are described in this document for all test simulations, but the data input and model output are not included as part of the printed document. An abbreviated Name file (NAM), DIS, BAS, Layer Property Flow (LPF), UZF1, SWR1, and external time series input datasets for test simulation 5 have been included in appendix 5, and can be used as a guide for developing MODFLOW datasets including the SWR1 Process. All five test simulations and this report are available online at *http://water.usgs.gov/software/*.

Test 1: Simple Reservoir Routing (Level Pool)

Test simulation 1 was developed to test the reservoir-routing approximation, and represents the outflow from a reservoir in response to an upstream inflow event. A MODFLOW model consisting of 1 row, 1 column, and 1 layer was developed. A horizontal grid spacing of 10 m was specified in the row and column directions, and an aquifer thickness of 10 m was specified. A surface-water-only model was simulated by setting ISWRONLY to 1. Although irrelevant to the simulation, a horizontal and vertical hydraulic conductivity of 1 m/sec, a specific storage value of 1 m^{-1}, and a specific yield of 0.2 were specified for the groundwater component.

A single reach was simulated. A SVAP table (IGEOTYPE=4) was used to represent the relation between stage and other reservoir parameters. Rainfall, evaporation, aquifer-reach exchanges, and uncontrolled lateral flows were not simulated, so arbitrary values were assigned to each defined reservoir stage for the surface area, wetted perimeter, and cross-sectional area. The SVAP table data (volume as a function of stage) and the stage-discharge relation used in test simulation 1 are summarized in table 5.

A transient simulation with 21 stress periods and a total simulation time of 210 minutes was simulated with a constant MODFLOW and SWR1 time step of 10 minutes. A specified inflow was applied to the reach using a list format and is summarized in table 6. The initial reach stage was set at 1.520 m.

SWR1 results for test simulation 1 were compared to a central-in-space backward-in-time version of the problem described in Bedient and Huber (1988). A comparison of flow results for SWR1 and Bedient and Huber (1988) is shown in figure 21A, along with the specified inflow. SWR1 results compare favorably to Bedient and Huber (1988), and demonstrate that SWR1 is

Table 5. Reservoir rating curve used in test simulation 1.

Elevation, in meters	Volume, in cubic meters	Discharge, in cubic meters per second
1.52	0	0
1.83	1,233.52	.42
2.13	2,467.03	.91
2.44	3,700.55	1.56
2.74	4,934.06	2.55
3.05	6,167.58	3.54
3.35	7,401.09	4.47
3.66	9,251.37	5.24
3.96	12,951.91	5.95
4.27	14,802.19	6.51
4.57	16,652.46	7.08
4.88	24,670.31	7.65
5.18	27,137.34	8.21

Table 6. Reservoir-specified inflow hydrograph used in test simulation 1.

Time, in minutes	Inflow, in cubic meters per second	Time, in minutes	Inflow, in cubic meters per second
10	0	120	4.53
20	1.70	130	3.40
30	3.40	140	2.27
40	5.10	150	1.13
50	6.80	160	0
60	8.50	170	0
70	10.19	180	0
80	9.06	190	0
90	7.93	200	0
100	6.80	210	0
110	5.66		

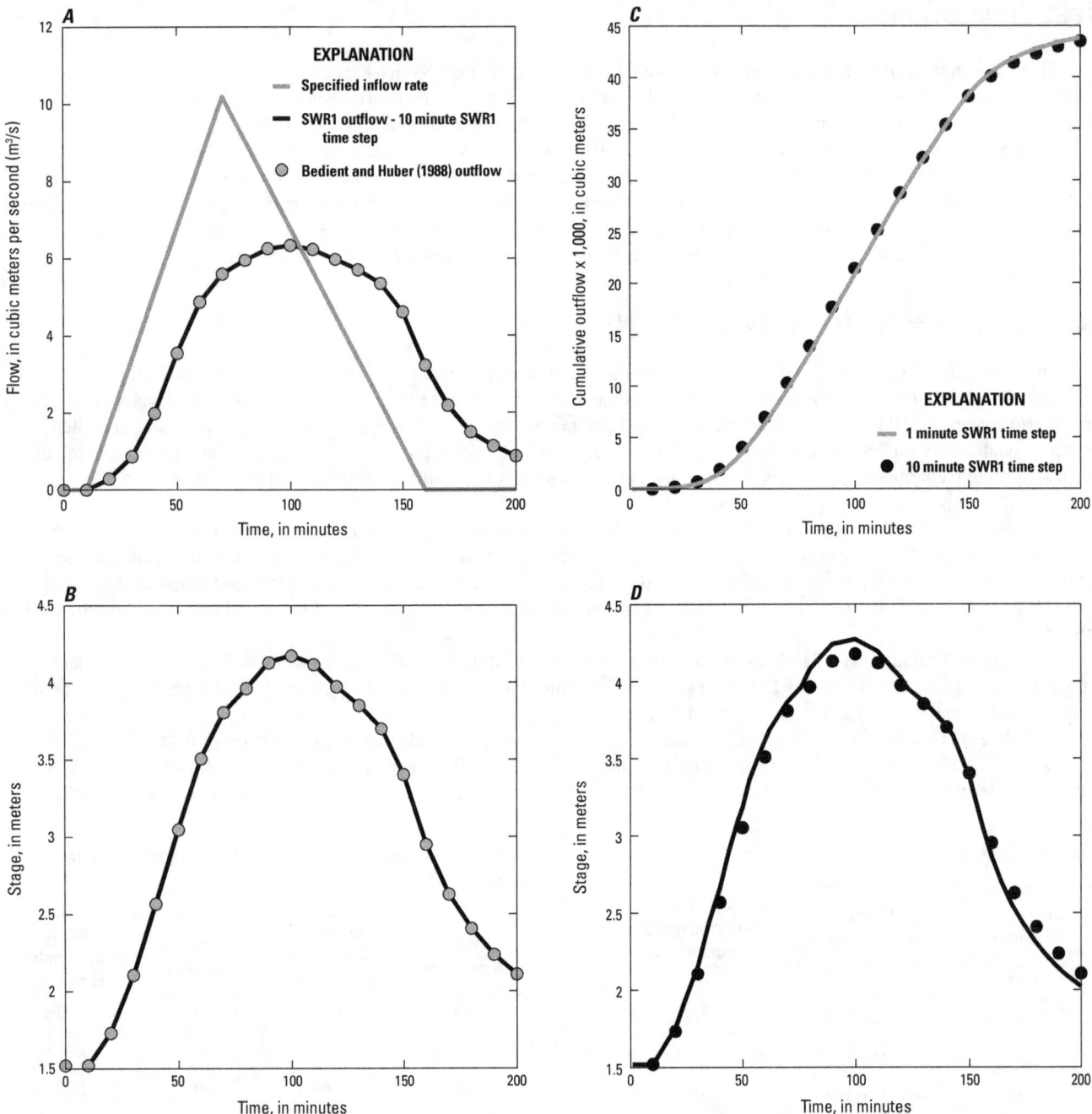

Figure 21.   Comparison of test simulation 1 SWR1 and Bedient and Huber (1988) results for, *A*, flow and *B*, stage. Comparison of *C*, cumulative outflow and *D*, stage results using a 1 and 10 minute SWR1 time step. The specified inflow applied in test simulation 1 is shown in *A*.

capable of accurately simulating reservoir routing. The difference between simulated stages and cumulative outflow volumes with a 1-minute SWR1 time step (10 SWR1 time steps in each MODFLOW time step) and a 10-minute SWR1 time step are shown in figure 21*B* and demonstrate that, for this problem, there are only slight differences in simulated results for the two SWR1 time steps. Differences between the two SWR1 time steps are a result of using a backward-in-time finite-difference scheme. No noticeable differences were observed when a central-in-time finite-difference scheme was used (not shown).

Test 2: Two-Dimensional Transient Shallow-Water Response To Constant Inflow

Test simulation 2 was developed to test diffusive-wave approximation routing for a transient two-dimensional problem with a constant inflow. This problem is conceptual, and represents the response of a wetland system to a constant upstream inflow. A MODFLOW model consisting of 11 rows, 11 columns, and 1 layer was developed. A horizontal grid spacing of 500 m in the row and column directions, a variable-top surface elevation ranging from 0.05 to 1.0 m, and an aquifer bottom elevation of -5 m were specified. The top-surface elevation, which is equivalent to the bathymetry of the shallow-water system, is shown in figure 22. A surface-water-only model was simulated by setting ISWRONLY to 1. Although irrelevant for the simulation, a horizontal and vertical hydraulic conductivity of 5 m/s, a specific storage value of 1 m^{-1}, and a specific yield of 0.2 were specified for the groundwater component.

A total of 121 model-based (IGEOTYPE=5) reaches were simulated. All reaches were assigned a Manning's n value of 0.30 m$^{1/3}$/s.

A transient simulation with 1 stress period, 5,040 time steps, and a total simulation time of 7 days was simulated with a constant MODFLOW and SWR1 time step of 2 minutes. Rainfall, evaporation, and aquifer-reach exchanges were not simulated. A constant lateral inflow rate of 23.570 m^3/s per inflow cell was assigned to the highest 11 cells (row 1) and a constant-stage value of 1.05 m was assigned to the lowest 11 cells (row 11). Constant lateral inflows were assigned using a two-dimensional array. Boundary condition locations are shown in figure 22. Initial stages were set at 2.05 m in all dynamic stage cells.

SWR1 results for test simulation 2 were evaluated at the three observation locations shown in figure 22, and were compared to results from SWIFT2D (Schaffranek, 2004). Comparisons of flow and stage results for SWR1 and SWIFT2D are shown in figure 23. SWR1 results compare favorably to SWIFT2D results, which demonstrates that SWR1 is capable of accurately simulating diffusive-wave approximation routing. For this case, the diffusive-wave approximation (maximum SWR1 Froude number = 0.049) compares favorably to a two-dimensional solution of the dynamic-wave approximation of the Saint-Venant equations simulated with SWIFT2D. The final water depth is 1.0 m in all SWR1 and SWIFT2D grid cells; the final water depth can be calculated directly from Manning's equation (eq. 6) using the specified Manning's roughness coefficient, inflow rate, and friction slope (equivalent to the bathymetric slope).

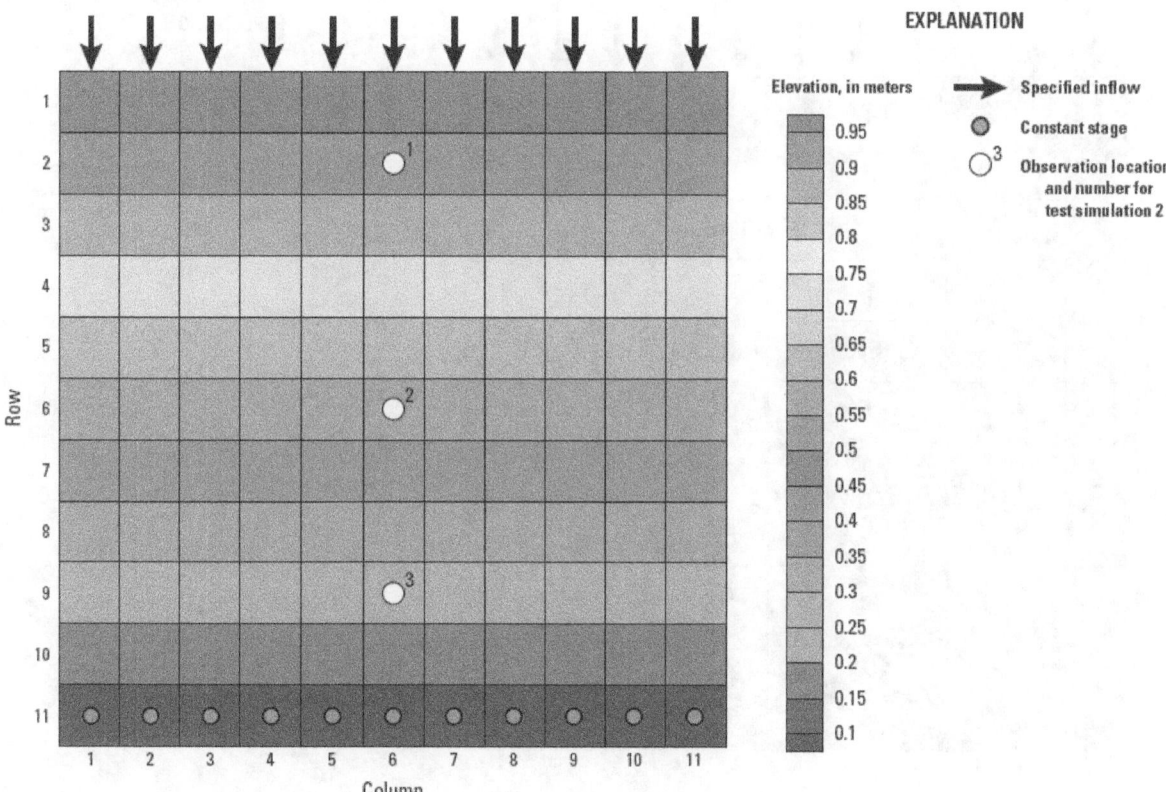

Figure 22.   Model grid showing the elevation of the top layer of the model (which can be considered to be the bathymetry of the shallow-water system), the location of boundary conditions, and the location of observation locations used in test simulation 2.

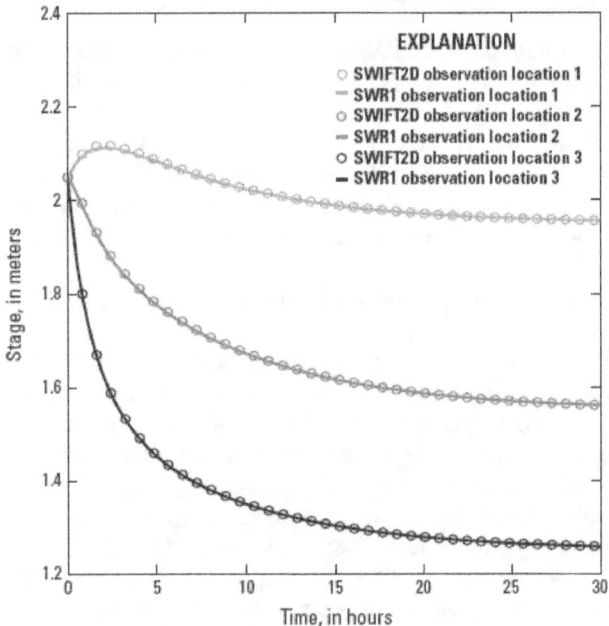

Figure 23. Comparison of test simulation 2 SWR1 and SWIFT2D stage results (Schaffranek, 2004).

Test 3: Two-Dimensional Transient Shallow-Water Response to Rainfall

Test simulation 3 was developed to test diffusive-wave approximation routing for a two-dimensional transient flow problem representing a wetland system with a specified-boundary inflow and spatially varying rainfall. The MODFLOW model setup is identical to that of test simulation 2. The top-surface elevation is also shown in figure 24. A surface-water-only model was simulated by setting ISWRONLY to 1.

A total of 121 model-based (IGEOTYPE=5) reaches were simulated. All reaches were assigned a Manning's n value of 0.03 m$^{1/3}$/s.

A transient simulation was constructed that incorporates 24 stress periods, each with 30 time steps, and 1 subsequent stress period having 4,320 time steps. The total simulation time for test simulation 3 was 7 days, with a constant MODFLOW and SWR1 time step of 2 minutes. Evaporation and aquifer-reach exchanges were not simulated. A time-varying rainfall rate was applied to the 16 cells in the central area of the active model domain, a time-varying lateral inflow was assigned to the highest 11 cells (row 1), and a constant stage value of 1.05 m was assigned to the lowest 11 cells (row 11). Lateral inflows and rainfall were assigned using two-dimensional arrays.

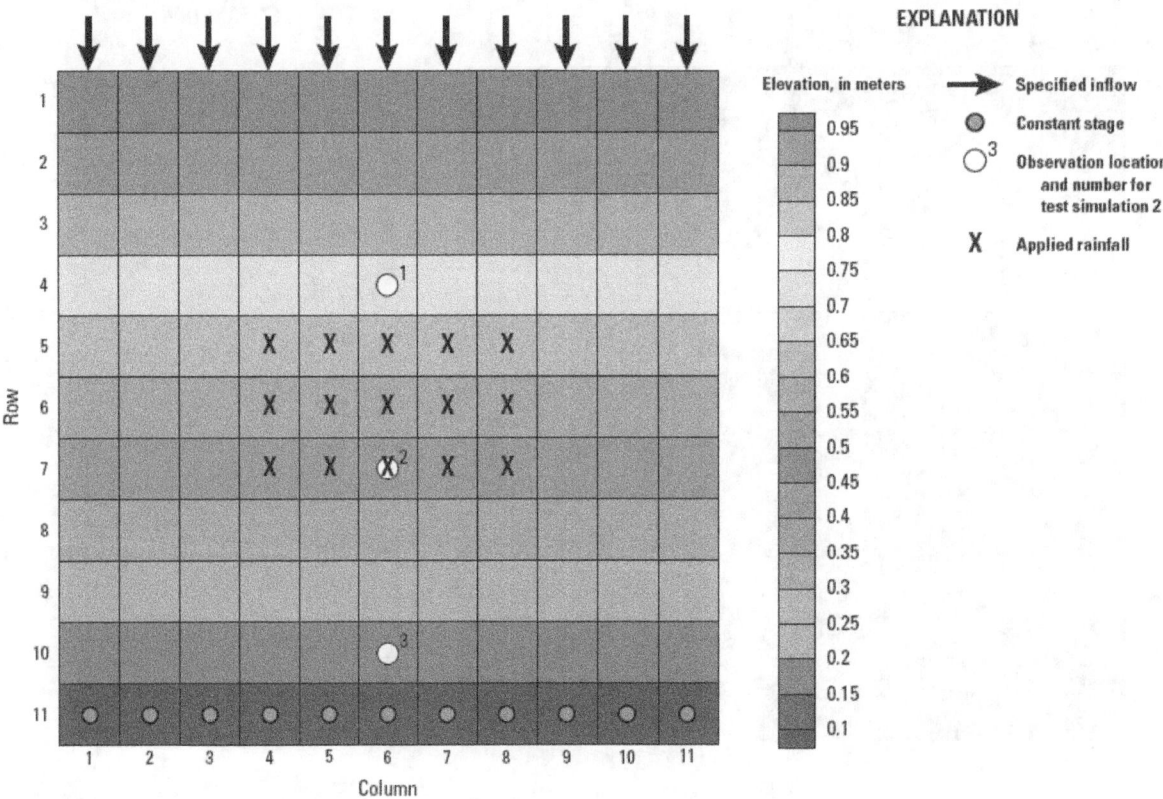

Figure 24. Model grid showing the elevation of the top of the model (which can be considered to be the bathymetry of the shallow-water system), the location of boundary conditions, and the location of observations used in test simulation 3.

Boundary-condition locations are shown in figure 24. The time-varying rainfall rate and volumetric inflow rate applied in test simulation 3 are shown in figure 25. Initial conditions were set at 0.05 m above the bottom of the reach in all dynamic stage cells.

SWR1 results for test simulation 3 were evaluated at the three observation locations shown in figure 24, and were compared to results from SWIFT2D (Schaffranek, 2004). A comparison of stage results for SWR1 and SWIFT2D is shown in figure 26. SWR1 results compare favorably to those of SWIFT2D and demonstrate that SWR1 is capable of accurately simulating transient diffusive-wave approximation routing with specified time-varying lateral and internal flow boundary conditions. Differences between SWR1 and SWIFT2D results are related to differences in the approaches used for wet and dry model grid cells. For this case, the diffusive-wave approximation (maximum SWR1 Froude number = 0.018) compares favorably to a two-dimensional solution of the dynamic-wave approximation of the Saint-Venant equations simulated with SWIFT2D.

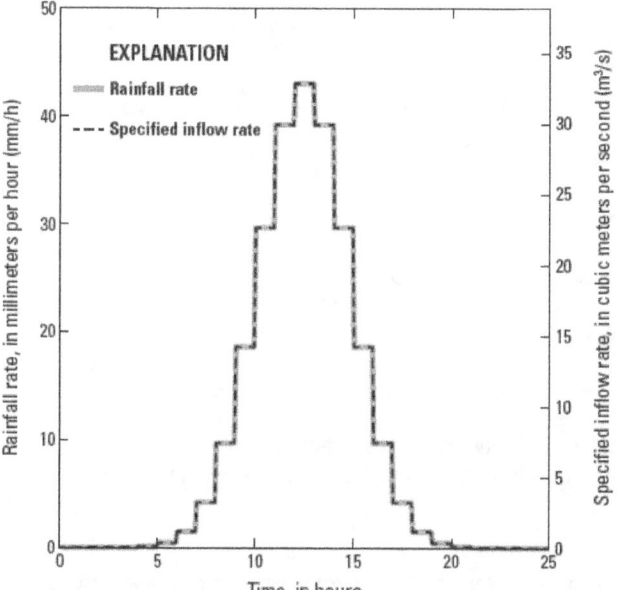

Figure 25. Rainfall rate and specified inflow applied in test simulation 3.

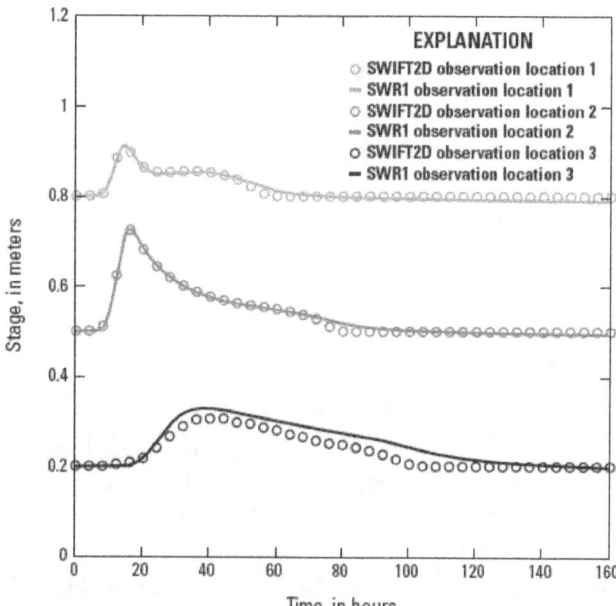

Figure 26. Comparison of test simulation 3 SWR1 and SWIFT2D stage results (Schaffranek, 2004).

Test 4: One-Dimensional Transient Looped Network

Test simulation 4 was developed to test diffusive-wave approximation routing for a one-dimensional transient flow problem for a looped network with multiple time-varying inflow conditions. A MODFLOW model consisting of 6 rows, 6 columns, and 1 layer was developed. A horizontal grid spacing of 1,000 m was specified in the row and column directions, and an aquifer thickness of 10 m was specified. A surface-water-only model was simulated by setting ISWRONLY to 1. Although irrelevant for the simulation, a horizontal and vertical hydraulic conductivity of 5 m/s, a specific storage value of 1 m^{-1}, and a specific yield of 0.2 were specified for the groundwater component.

A total of 18 trapezoidal (IGEOTYPE=2) reaches were simulated and are shown in figure 27. Reach parameters are summarized in table 7 and cross-section data are summarized in table 8. A constant bottom elevation of 0.0 m was specified for all cross sections but was vertically adjusted using a specified vertical cross-section offset (GZSHIFT) for each reach (table 7). This approach reduces the number of cross sections specified in the SWR1 input file, 3 in test simulation 4, and could be used in SWR1 input files that have reaches with a limited number of station-elevation geometries but different bottom elevations. All reaches were assigned a Manning's n value of 0.03 m$^{1/3}$/s. Two fixed crest weirs were simulated in reaches 14 and 18 (fig. 27). Structure characteristics are summarized in table 9. A SWR1 submergence exponent (STRCD3) of zero was used to reduce the discrepancies between the weir equations implemented in the SWR1 Process and HEC-RAS.

A transient simulation was formulated consisting of 24 stress periods, each with 2 time steps, and 1 subsequent stress period having 288 time steps. The total simulation time for test simulation 4 was 7 days, with a constant MODFLOW and SWR1 time step of 30 minutes. Evaporation and aquifer-reach exchanges were not simulated. Time-varying lateral inflows were applied

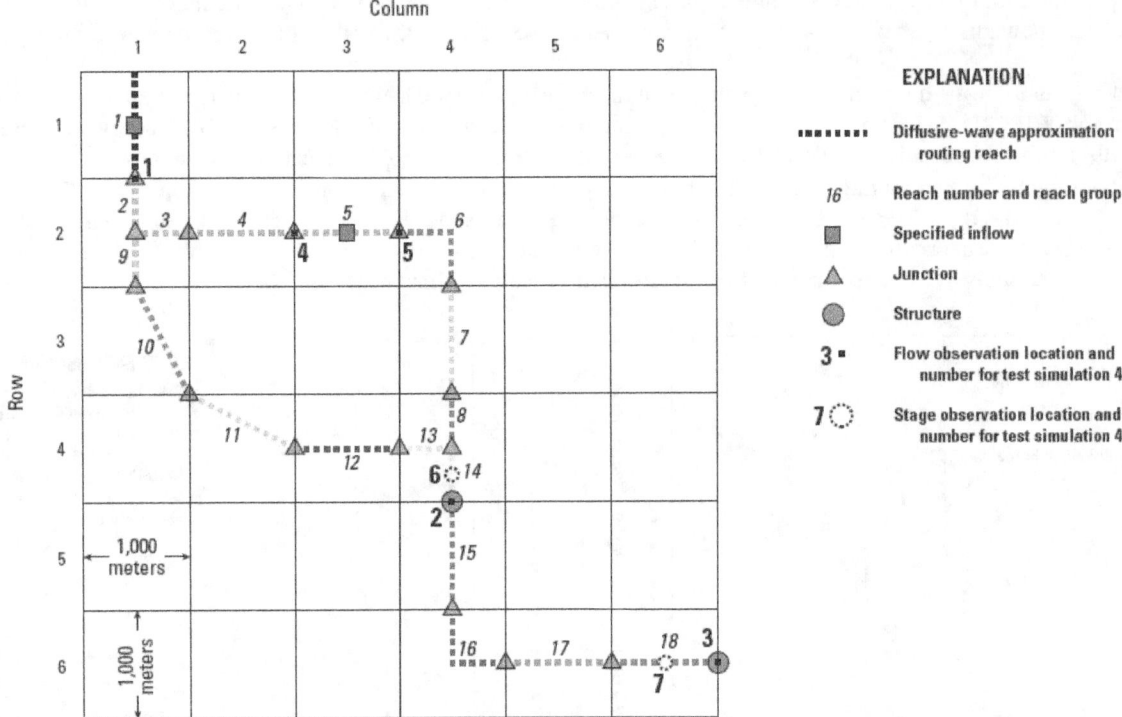

Figure 27. Model grid showing the location of one-dimensional looped network reaches, junctions, structures, boundary conditions, and observations used in test simulation 4. Reach numbers are equivalent to the reach group number in test simulation 4.

in reaches 1 and 5. External time series files were used to assign time-varying lateral inflows. Boundary-condition locations are shown in figure 27. The structure in reach 18 is used to represent stage-dependent discharge from the model. The time-varying lateral inflow rates applied in test simulation 4 are shown in figure 28. Initial stages were set at 3.0 m in reaches 1 through 14, and 2.0 m in reaches 15 through 18.

SWR1 results for test simulation 4 were evaluated at the seven observation locations shown in figure 27, and were compared to results from HEC-RAS (U.S. Army Corps of Engineers, 2008). Comparisons of stage results for SWR1 and HEC-RAS are shown in figure 29. SWR1 results compare favorably to HEC-RAS and demonstrate that SWR1 is capable of accurately simulating transient one-dimensional surface-water routing with time-varying lateral flow boundary conditions and fixed control structures. Results also show that SWR1 accurately simulates flow reversals that occur in reach 5 (fig 29B) that result from differences in the magnitude and timing of lateral flow entering the model at reaches 1 and 5. For this case, the diffusive-wave approximation (maximum SWR1 Froude number = 0.19) compares favorably to a one-dimensional solution of the dynamic-wave approximation of the Saint-Venant equations simulated with HEC-RAS.

Test 5: Aquifer-Reach Interaction

Test simulation 5 was developed to test aquifer-reach interaction in a complex synthetic network that includes examples of constant- and dynamic-stage reaches, all of the reach-routing types, all geometry types available in SWR1, different conductance options, and a number of different structure types. An external time series file is used to specify stage for constant-stage reaches. This test problem represents a managed network of surface-water features receiving dynamic, rainfall-driven runoff and interacting with an underlying groundwater system. Abbreviated listings of the NAM, DIS, BAS, LPF, UZF1, SWR1, and external time series input datasets for test simulation 5 are included in appendix 5. A MODFLOW model consisting of 6 rows, 6 columns, and 2 layers was developed. The groundwater model was defined using a horizontal grid spacing of 304.8 m in the row and column directions, a variable top-surface elevation ranging from 0.305 to 1.824 m, and an aquifer bottom elevation of -20 m. The top surface elevation of the model is shown in figure 30 and listed in appendix 5. The lowest elevation is in row 3 column 5, and row 4 column 5, and corresponds to the location of a two-dimensional surface-water body (fig. 31). The bottom of layer 1 was defined as having an elevation of 0.250 m. A horizontal hydraulic conductivity ranging from 5 to 10 m/d (app. 5), a specific

Table 7.   Test simulation 4 SWR1 reach parameters.

Reach	Reach length, in meters	Initial stage, in meters	Cross section number[1]	Vertical cross-section offset, in meters
1	1,000	3.0	1	2.0
2	500	3.0	1	1.83
3	500	3.0	2	1.83
4	1,000	3.0	2	1.67
5	1,000	3.0	2	1.50
6	1,000	3.0	2	1.33
7	1,000	3.0	2	1.17
8	500	3.0	2	1.0
9	500	3.0	1	1.83
10	1,118	3.0	1	1.65
11	1,118	3.0	1	1.41
12	1,000	3.0	1	1.17
13	500	3.0	1	1.0
14	500	3.0	3	1.0
15	1,000	2.0	3	.75
16	1,000	2.0	3	.50
17	1,000	2.0	3	.25
18	1,000	2.0	3	.0

[1]Referenced cross-section number shown in table 8.

Table 8.   Test simulation 4 SWR1 reach geometry parameters.

Cross-section number	Width, in meters	Bottom elevation, in meters	Side slope, in meters per meter	Manning's roughness coefficient, in seconds per meter$^{1/3}$
1	10	0.0	1.0	0.03
2	20	.0	1.0	.03
3	15	.0	1.0	.03

Table 9.   Test simulation 4 SWR1 structure parameters.

Reach	Structure type	Invert elevation, in meters	Structure width, in meters	Structure submergence exponent
14	Weir	3.0	10.0	0.0
18	Weir	2.0	10.0	0.0

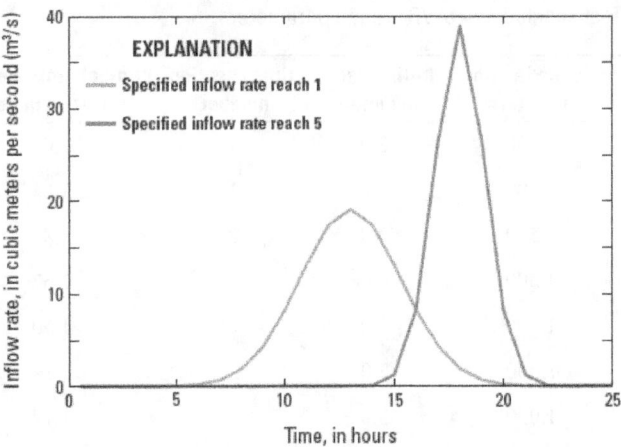

Figure 28. Specified inflow applied to reaches 1 and 5 in test simulation 4.

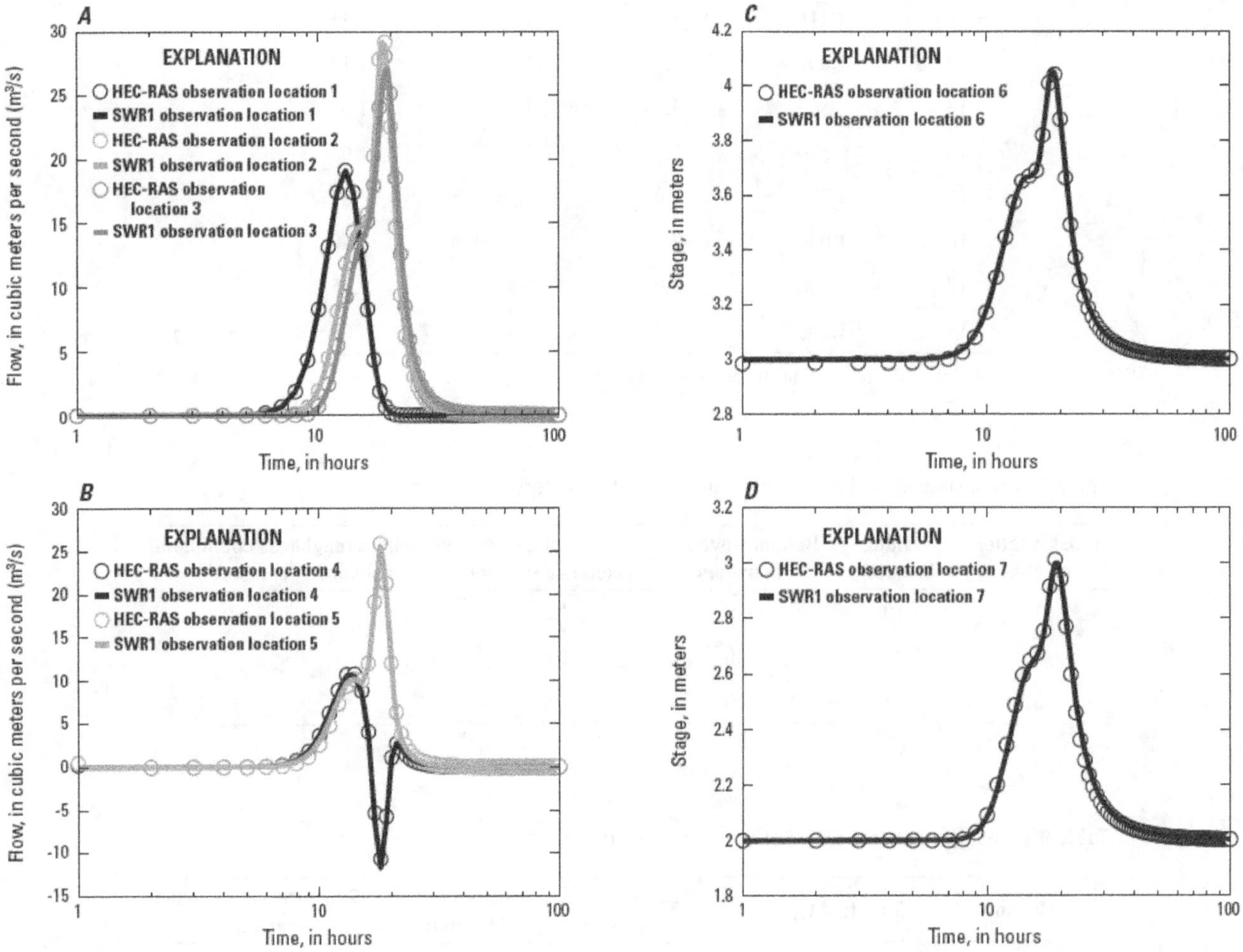

Figure 29. Comparison of test simulation 4 SWR1 and Hydrologic Engineering Center River Analysis System (HEC-RAS) (U.S. Army Corps of Engineers, 2008) results for discharge at, *A*, observation locations 1, 2, and 3; *B*, observation locations 4 and 5; and stage at *C*, observation location 6, and *D*, observation location 7. Locations shown in figure 27.   Negative discharge in *B*, indicates flow in the upstream direction, from reach 5 toward reach 4.

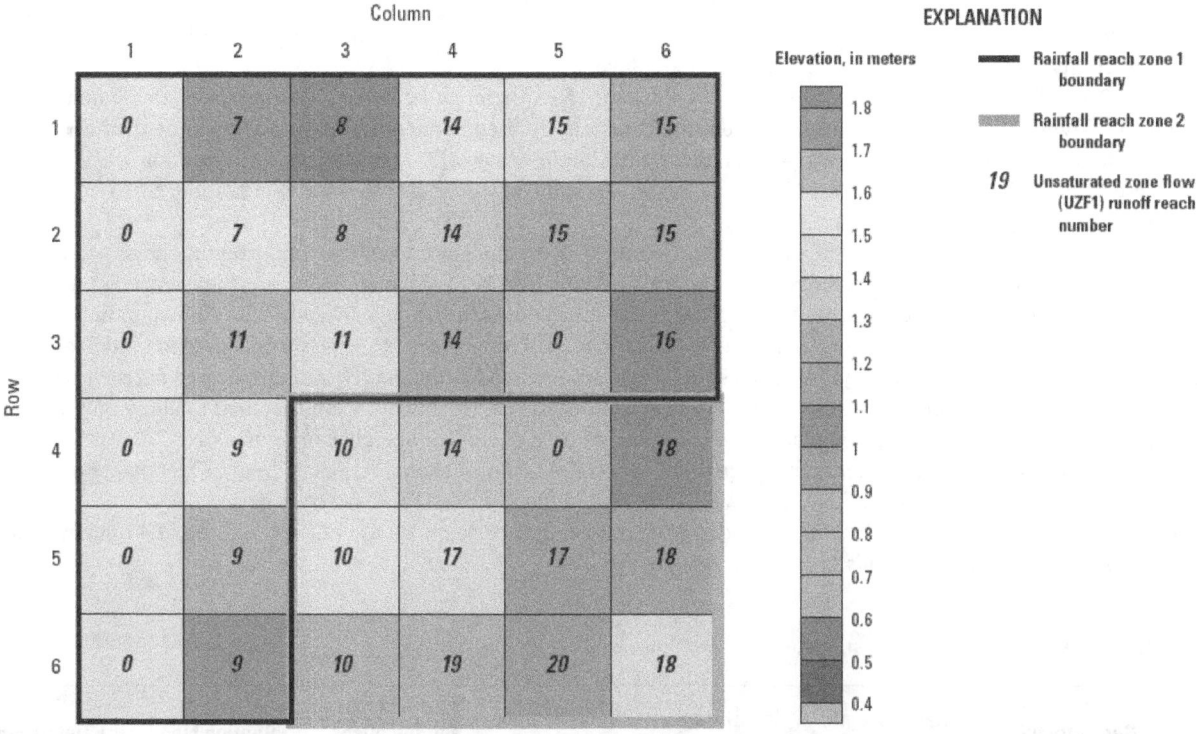

Figure 30. Model grid showing the model topography (m), surface-water runoff contributing areas for reaches shown in figure 31, and rainfall zones used in test simulation 5. Unsaturated zone flow and surface-water runoff are not simulated in cells with unsaturated zone flow runoff reach numbers set equal to zero.

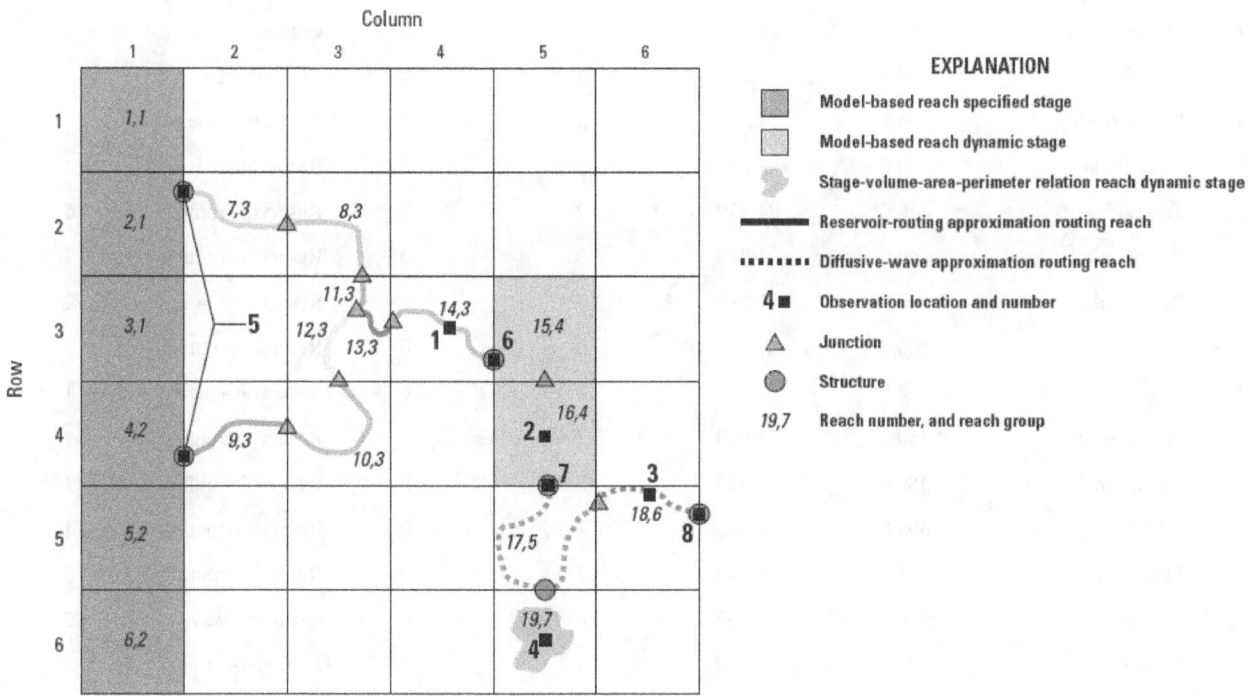

Figure 31. Model grid showing the location of one- and two-dimensional network reaches, junctions, structures, boundary conditions, and observations used in test simulation 5.

storage value of 1×10^{-5} m^{-1}, and a specific yield of 0.20 was specified for model layer 1. A horizontal hydraulic conductivity of 5 m/d, a specific storage value of 1×10^{-5} m^{-1}, and a specific yield of 0.20 were specified for model layer 2. A horizontal-to-vertical hydraulic conductivity ratio of 10.0 was applied to both model layers.

Unsaturated zone flow was simulated using the UZF1 Package (Niswonger and others, 2006) in the cells with non-zero numbers shown in figure 30. The saturated vertical hydraulic conductivity used in the unsaturated zone was defined using groundwater data in the LPF Package (IUZOPT=2). A total of 20 trailing waves (NTRAIL2) and 20 infiltration wave sets (NSETS2) were simulated. A constant Brooks-Corey epsilon (EPS) value of 3.5, a saturated water content (THTS) of 0.25, and an initial water content (THTI) of 0.15 were specified.

A total of 19 reaches were simulated and are shown in figure 31. Rectangular, trapezoidal, and irregular cross sections are included in the model along with model-based reaches and a SVAP table reach. Reach parameters are summarized in table 10 and cross-section data are summarized in table 11. Irregular cross-section data for reaches 17 and 18 are shown in figure 32 and SVAP table data for reach 19 are summarized in table 12. Reaches 1 through 16 and 19 were connected to model layer 1. Reaches 17 and 18 intersect both model layers. Seven structures were included in the model, and are located in reaches 2, 4, 14, 16, 18, and 19. Structure details are summarized in table 13. Structures in reaches 2 and 4 were specified to allow only downstream flow (ISTRDIR=1). Reach stages in reaches 14 and 19 were used to control operable structures (ISTRORCH) in these reaches when simulated stages were greater than or equal to the specified control elevation, respectively. Operable structures were closed (STRVAL) at the beginning of the first SWR1 time step in the first MODFLOW stress period.

A transient simulation with 367 daily stress periods was formulated. A constant MODFLOW time step of 4 hours was used. SWR1 time steps ranging from 30 seconds to 30 minutes were used.

Table 10. Test simulation 5 SWR1 reach parameters.

Reach	Geometry type	Reach length, in meters	Initial stage, in meters	Cross section number[1]	Vertical cross-section offset, in meters	Routing type	Rainfall zone
1	Model-based	304.8	1.537	1	0.0	Constant stage	1
2	Model-based	304.8	1.537	1	.0	Constant stage	1
3	Model-based	304.8	1.537	1	.0	Constant stage	1
4	Model-based	304.8	1.537	1	.0	Constant stage	1
5	Model-based	304.8	1.537	1	.0	Constant stage	1
6	Model-based	304.8	1.537	1	.0	Constant stage	1
7	Trapezoidal	350.0	1.219	2	.0	Reservoir-routing	1
8	Trapezoidal	350.0	1.219	2	.0	Reservoir-routing	1
9	Rectangular	340.0	1.219	3	.0	Reservoir-routing	1
10	Rectangular	460.0	1.219	3	.0	Reservoir-routing	2
11	Trapezoidal	100.0	1.219	2	.0	Reservoir-routing	1
12	Trapezoidal	250.0	1.219	2	.0	Reservoir-routing	1
13	Trapezoidal	150.0	1.219	2	.0	Reservoir-routing	1
14	Trapezoidal	380.0	1.219	2	.0	Reservoir-routing	1
15	Model-based	304.8	1.000	4	.0	Reservoir-routing	1
16	Model-based	304.8	1.000	4	.0	Reservoir-routing	2
17	Irregular	790.0	.609	6	.0	Diffusive-wave	2
18	Irregular	325.0	.609	7	.0	Diffusive-wave	2
19	Stage-area-volume-perimeter table	150.4	1.000	5	.0	Reservoir-routing	2

[1]Referenced cross-section number shown in table 11.

Table 11.   Test simulation 5 SWR1 reach geometry parameters.

Cross-section number	Width, in meters	Bottom elevation, in meters	Side slope, in meters per meter	Manning's roughness coefficient, in seconds per meter$^{1/3}$	Conductance type	Conductance, in square meters per day	Reach leakance, per day	Reach leakance length, in meters
1	—	—	—	0.25	Conductance	9.20×10^4	—	—
2	30	0.3048	1.0	.03	Reach bed	—	2.50×10^{-1}	—
3	30	.3048	—	.03	Reach bed	—	2.50×10^{-1}	—
4	—	—	—	.25	Reach bed	—	2.00×10^{-2}	—
5	—	—	—	.05	Reach bed	—	1.00×10^{-2}	—
6	—	—	—	.10	Reach bed/ aquifer	—	1.00×10^{-1}	63.50
7	—	—	—	.03	Reach bed	—	5.00×10^{-1}	—

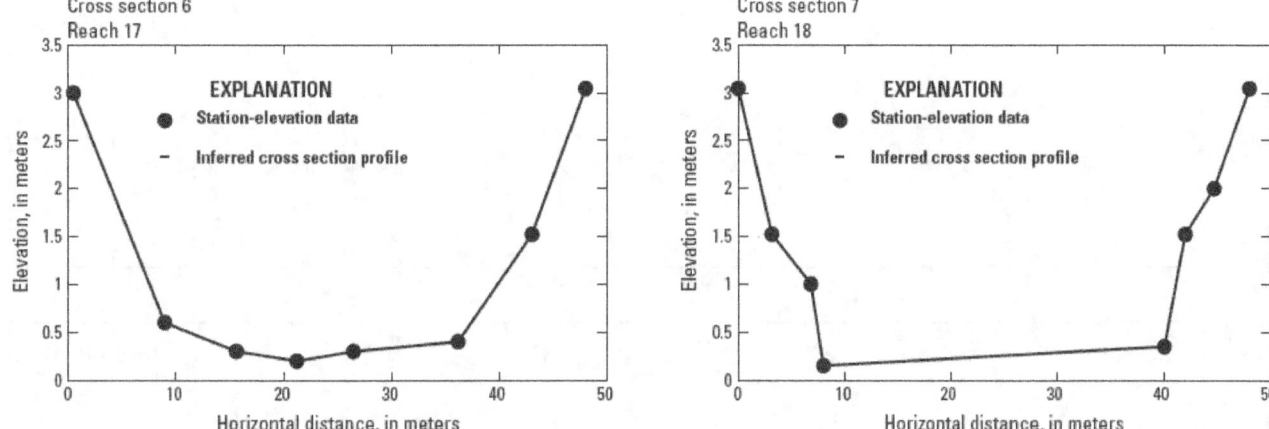

Figure 32.   Irregular cross sections used in test simulation 5 for reaches 17 and 18.

Rainfall data from two stations (fig. 33) were applied to the rainfall zones shown in figure 30. Constant evaporation and evapotranspiration rates of 3.75×10^{-3} and 2.63×10^{-3} m/d, respectively, were used throughout the entire simulation period. The UZF1 Package was used to calculate evapotranspiration from the unsaturated zone and groundwater system and calculate infiltration, runoff, and groundwater recharge in model cells having a non-zero value in figure 30. A constant evapotranspiration extinction depth and water content of 0.25 m and 0.05 were specified, respectively. Overland runoff, composed of excess rainfall and groundwater discharge to the land surface, was calculated by the UZF1 Package and routed (IRUNFLG>0) to the reach numbers identified in figure 30 (reaches are shown in fig. 31). An average undulation depth (SURFDEP) of 0.5 m was specified. Rainfall was applied directly to all reaches using a two-dimensional array. Evaporation was simulated in all reaches and specified using a list format. Model-based reaches were assigned a 0.25-m extinction depth (GETXTD). A constant lateral inflow of 0.0 m³/d was specified for all reaches, and was used throughout the entire simulation period.

Model-based reaches in column 1 were simulated as constant-stage reaches using the boundary stage data shown in figure 33C. A general head boundary with a groundwater head of 1.534 m and a conductance of 1×10^4 m²/d was applied to all groundwater cells in column 1 of model layer 1. General head boundaries, each with a constant conductance of 1×10^4 m²/d and groundwater head of 0.510 and 0.610 m, were applied in row 5 column 6 in model layers 1 and 2, respectively, to allow groundwater outflow from model cells beneath reach 18. Initial stage in the dynamic-stage reaches was set at 1.22 m in reaches 7 through 14, 1.00 m in reaches 15 and 16, 0.61 m in reaches 17 and 18, and 1.00 m in reach 19. Variable initial groundwater heads were specified in both model layers and are listed in appendix 5.

SWR1 results for test simulation 5 were evaluated at the eight observation locations shown in figure 31. Simulated stage and flow results at selected observation locations are shown in figure 34A-B. Simulated groundwater heads at observation

Table 12.　Test simulation 5 SWR1 reservoir SVAP table for reach 19 representing a conic lake smaller than the model grid cell.

Elevation, in meters	Volume, in cubic meters	Wetted perimeter, in meters	Surface area, in square meters	Cross-sectional area, in square meters
0.3048	0	0	0	0
.5096	78	2,280	1,140	3.07
.7144	623	9,121	4,560	6.14
.9192	2,101	20,523	10,261	9.22
1.1240	4,981	36,485	18,242	12.29
1.1241	4,991	111,146	92,903	12.29
4.5721	325,320	111,146	92,903	64.01

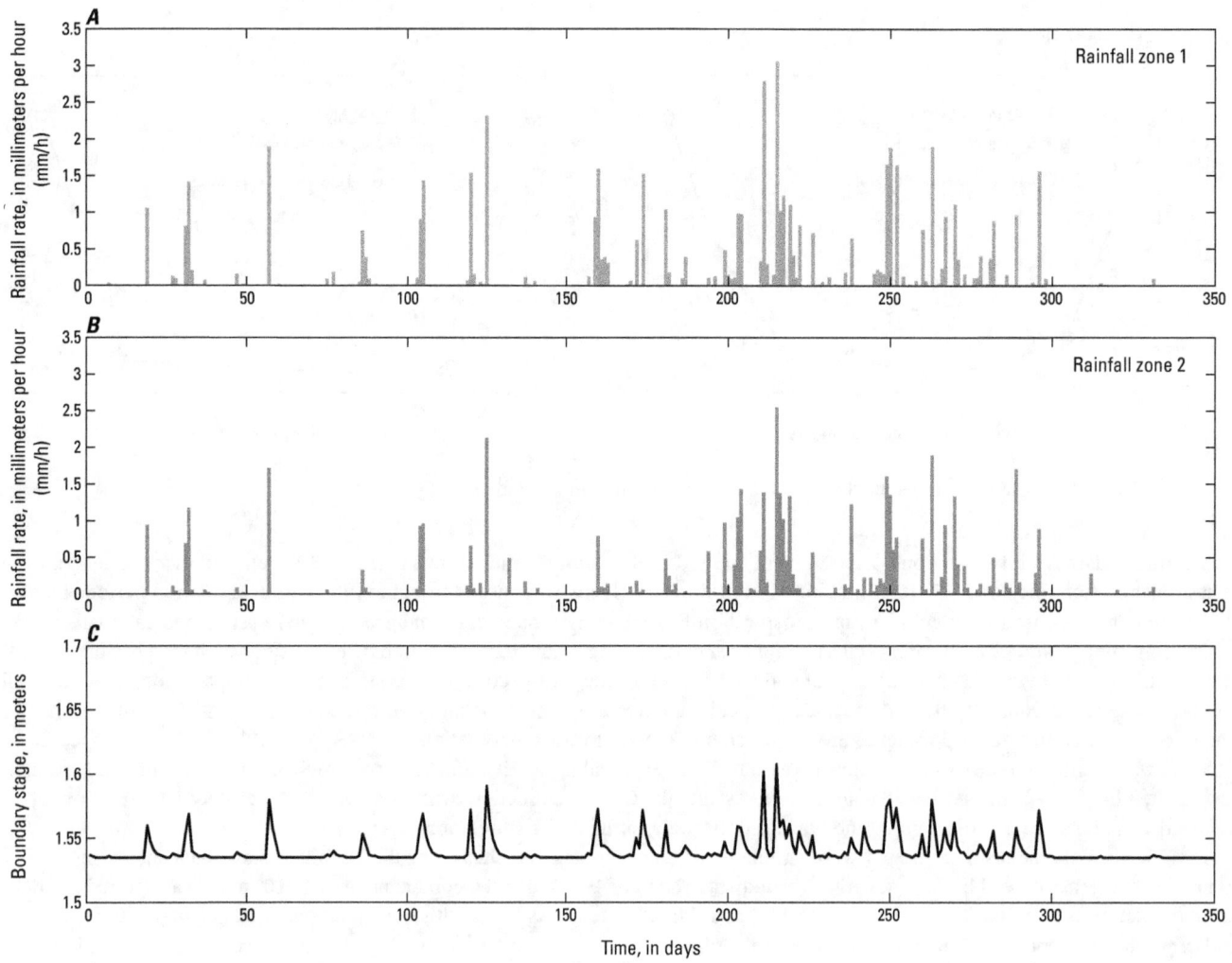

Figure 33.　Daily rainfall rate (mm/h) used in *A,* rainfall zone 1, *B,* rainfall zone 2, and *C,* specified stage values used in SWR1 reaches 1 through 6 in test simulation 5.

Table 13.   Test simulation 5 SWR1 structure parameters.

Reach	Structure type	Invert elevation, in meters	Structure width, in meters	Structure discharge coefficient	Structure submergence exponent	Maximum gate opening, in meters	Gate opening rate, in meters per second	Control elevation, in meters	Maximum discharge rate, in cubic meters per second
2	Weir	1.535	45.0	0.61	0.50	—	—	—	—
4	Weir	1.535	45.0	.61	.50	—	—	—	—
14	Gated spillway	1.000	30.0	.61/.61[1]	.50	0.25	0.00125	1.25	—
14	Pump	—	—	—	—	—	.00521[2]	1.75	0.25
16	Weir	1.000	40.0	.61	.50	—	—	—	—
18	Weir	.500	45.0	.61	.50	—	—	—	—
19	Overflow weir	.45	10.0	.61	.50	.25	.00125	.75	—

[1]Weir/orifice discharge coefficient.

[2]Gate opening rate for pump has units of cubic meters per second.

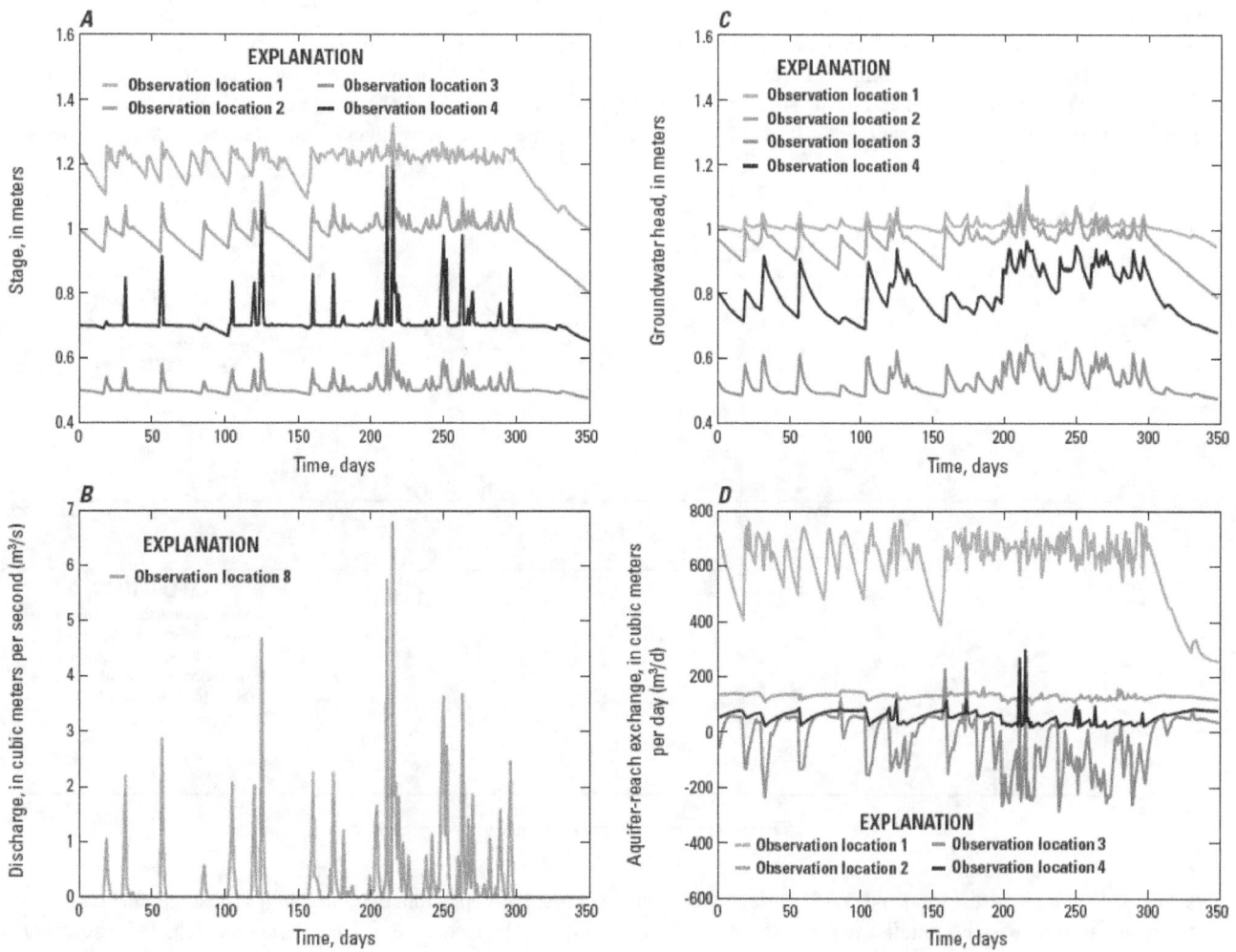

Figure 34.   *A*, Simulated test simulation 5 SWR1 results for stage at observation locations 1–4, *B*, discharge at downstream observation location 8, *C*, groundwater head at observation locations 1–4, and *D*, aquifer-reach exchange at observation locations 1–4 shown in figure 31.   Note: negative aquifer-reach exchange in *D*, indicates flow from the aquifer to the reach.

locations 1 through 4 are shown in figure 34*C*. Simulated aquifer-reach exchange for the observation locations shown in figure 31 are shown in figure 34*D*. The total simulated UZF1 runoff, inflow to reach group 3 from constant stage reaches 1 through 6, and downstream discharge from reach 18 are shown in figure 35*A*. The simulated flow duration at observation locations 5 through 8 and the total UZF1 runoff are shown in figure 35*B*. The simulated SWR1 water budget for test simulation 5 at the end of stress period 200 is shown in figure 36. Results shown in figures 34 and 35 represent typical analyses that can be performed on MODFLOW models using the SWR1 Process to evaluate conjunctive use of surface water and groundwater. The total run-time for this one-year transient simulation period was less than one minute (on a 2.67 GHz Intel i7 central processing unit (CPU) with 4 gigabytes (GB) of random access memory (RAM) running the Windows XP x64 operating system) and indicates that it would be practical to extend the simulation period to include several years to decades of rainfall and evapotranspiration data.

Simulated stages at the end of stress period 200 were extracted along with calculated reach conductances and groundwater levels to develop a model with a constant stage SWR1 dataset that could be compared to an equivalent model using the MOD-FLOW RIV Package. The stages specified in the SWR1 and RIV Package datasets are summarized in table 14. Reaches that intersect both layers, reaches 17 and 18, were simulated in the MODFLOW RIV Package using river boundaries in both model layers and the appropriate layer conductance from the transient model. The conductances specified in the constant stage RIV model are summarized in table 14. The specified-reach stage and aquifer reach options summarized in table 11 were used by the SWR1 Process to internally calculate the conductance applied to reaches.

A steady-state simulation using constant reach stages with a total simulation time of 1 day, using 1 stress period having 1 time step, was formulated. SWR1 time steps in the constant-reach-stage SWR1 dataset were identical to MODFLOW time steps. The constant-reach stage model was run using average rainfall rates over the 367-day transient simulation period (0.153 and 0.127 mm/h for rainfall zones 1 and 2, respectively) applied in the UZF1 Package. Rainfall and evaporation were not applied to SWR1 reaches in the constant-reach-stage SWR1 dataset. All other model parameters and boundary conditions used in the constant-reach-stage model were identical to those of the transient model. The constant-reach stage-model with the constant-reach-stage SWR1 dataset compare favorably to the equivalent RIV Package. Simulated aquifer-reach exchanges for the SWR1 and RIV Packages at the end of the simulation are summarized in table 14. Discrepancies between the two results are ±0.0 percent, which demonstrates that the SWR1 Process can be configured in a manner that is consistent with the RIV Package.

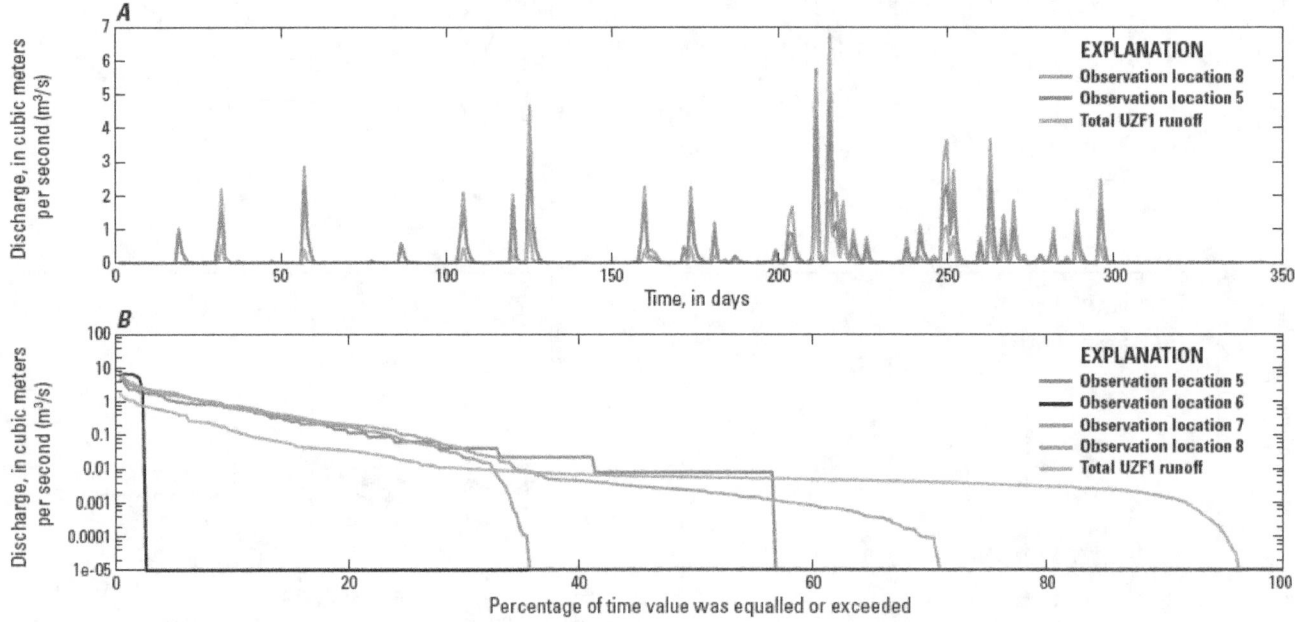

Figure 35. *A*, Simulated test simulation 5 SWR1 surface-water inflow to reach group 3 from reaches 2 and 4 (observation location 5), total unsaturated zone (UZF1) runoff, and downstream discharge at observation location 8. *B*, Flow duration results for observation locations 5–8 shown in figure 31 and UZF1 runoff.

```
                 VOLUMETRIC SURFACE WATER BUDGET FOR ENTIRE MODEL
                  AT END OF TIME STEP     1 IN STRESS PERIOD  200
-------------------------------------------------------------------------------

     CUMULATIVE VOLUMES      L**3      RATES FOR THIS TIME STEP     L**3/T
     ------------------                -----------------------

            IN:                               IN:
            ---                               ---
     LATERAL FLOW  =          0.0000    LATERAL FLOW   =         0.0000
     UNSAT. RUNOFF =     418263.0312    UNSAT. RUNOFF  =      2563.8252
         RAINFALL  =     176870.5469        RAINFALL   =      1190.2156
      EVAPORATION  =          0.0000     EVAPORATION   =         0.0000
     REACH-AQ FLOW =      93783.7812    REACH-AQ FLOW  =       982.7893
     EXTERNAL FLOW =          0.0000    EXTERNAL FLOW  =         0.0000
     BOUNDARY FLOW =          0.0000    BOUNDARY FLOW  =         0.0000
     CONSTANT FLOW =    2829967.0000    CONSTANT FLOW  =         0.0000
          STORAGE  =     981206.8125         STORAGE   =     49154.7461

        TOTAL IN  =    4500091.0000        TOTAL IN   =     53891.5781

           OUT:                              OUT:
           ----                              ----
     LATERAL FLOW  =          0.0000    LATERAL FLOW   =         0.0000
     UNSAT. RUNOFF =          0.0000    UNSAT. RUNOFF  =         0.0000
         RAINFALL  =          0.0000        RAINFALL   =         0.0000
      EVAPORATION  =     476358.3438     EVAPORATION   =      3184.9158
     REACH-AQ FLOW =     277166.2500    REACH-AQ FLOW  =      1018.6852
     EXTERNAL FLOW =          0.0000    EXTERNAL FLOW  =         0.0000
     BOUNDARY FLOW =    2613033.5000    BOUNDARY FLOW  =     30730.6914
     CONSTANT FLOW =     153084.6875    CONSTANT FLOW  =     18989.1191
          STORAGE  =     979448.0625         STORAGE   =         0.0000

        TOTAL OUT =    4499091.0000        TOTAL OUT  =     53923.4141

        IN - OUT  =       1000.0000        IN - OUT   =       -31.8359

     PERCENT DISCREPANCY =       0.02    PERCENT DISCREPANCY =      -0.06


              DISCREPANCY BETWEEN MODFLOW AND SWR1 AQUIFER-REACH
                      TERMS FOR THE ENTIRE MODEL
-------------------------------------------------------------------------------
     CUMULATIVE VOLUMES      L**3      RATES FOR THIS TIME STEP     L**3/T
     ------------------                -----------------------
            MODFLOW =      183418.3125              MODFLOW =       35.5464
               SWR1 =      183382.4688                 SWR1 =       35.8959
  PERCENT DISCREPANCY =        -0.02    PERCENT DISCREPANCY =        0.98
SWR1 INACTIVE MODFLOW =        0.0000  SWR1 INACTIVE MODFLOW =        0.0000
-------------------------------------------------------------------------------
        PERCENT DISCREPANCY = 100 x (SWR1 - MODFLOW) / MODFLOW
```

Figure 36. Simulated cumulative and incremental surface-water budget for test simulation 5 at the end of time step 1 in stress period 200. The discrepancy between MODFLOW and SWR1 aquifer-reach exchanges is also shown.

Table 14. Test simulation 5 specified stage and reach parameters used for comparison with the MODFLOW River Package, and simulated results using the SWR1 Process and RIV Package.

Reach	Layer	Specified stage, in meters	Conductance, in square meters per day	Bottom elevation, in meters	Aquifer-reach exchange, in cubic meters per day		Percent discrepancy
					Constant-stage SWR1	Steady-state RIV	
1	1	1.539	92000	1.530	50.38	50.38	0.00
2	1	1.539	92000	1.530	68.81	68.81	.00
3	1	1.539	92000	1.530	60.87	60.87	.00
4	1	1.539	92000	1.530	69.50	69.50	.00
5	1	1.539	92000	1.530	59.75	59.75	.00
6	1	1.539	92000	1.530	50.41	50.41	.00
7	1	1.219	2926.7	0.3048	-159.1	-159.1	.00
8	1	1.219	2951.5	0.3048	-189.9	-189.9	.00
9	1	1.219	2757.3	0.3048	-144.1	-144.1	.00
10	1	1.219	3707.4	0.3048	71.29	71.29	.00
11	1	1.219	829.14	0.3048	11.91	11.91	.00
12	1	1.219	2072.8	0.3048	29.78	29.78	.00
13	1	1.219	1243.7	0.3048	17.87	17.87	.00
14	1	1.219	3097.0	0.3048	622.8	622.8	.00
15	1	1.006	1858.1	0.3048	10.64	10.64	.00
16	1	1.006	1858.1	0.3048	39.56	39.56	.00
17	1	0.5127	1266.4	0.2000	-154.9	-154.9	.00
17	2	0.5127	187.96	0.2000	-26.02	-26.02	.00
18	1	0.5080	2912.7	0.1500	-91.45	-91.45	.00
18	2	0.5080	2631.2	0.1500	-108.0	-108.0	.00
19	1	0.7008	364.85	0.3048	-65.56	-65.56	.00

Summary

The Surface-Water Routing (SWR1) Process developed for MODFLOW is capable of simulating one-dimensional and two-dimensional routing of surface-water flow. Simple reservoir-routing, diffusive-wave approximation routing, and constant-stage reaches can be simulated and included in the same model. The approach used to represent surface-water reaches allows a variety of geometric forms to be represented in the SWR1 Process. Rectangular, trapezoidal, and irregular cross sections can simulate one-dimensional surface-water features such as streams and canals. Two-dimensional surface-water features such as lakes and wetlands can be simulated using specified stage-volume-area-perimeter (SVAP) tables or reaches that cover an entire finite-difference grid cell (model-based reaches). Specified SVAP tables can be used to represent surface-water features that are smaller than the finite-difference grid cell or cannot be accurately represented using the defined topography of the grid cell.

Physical properties of surface-water features (reach geometry, reach-bed roughness, aquifer-reach interaction parameters, structure geometry, and structure control criteria), rainfall, evaporation, and specified lateral flows are assigned to SWR1 reaches. Several reaches can be within a single cell but a single reach cannot interact with more than one row and column location. Reach groups can be composed of individual reaches or multiple reaches. The SWR1 Process also can be used with the MODFLOW Unsaturated Zone Flow (UZF1) Package to permit dynamic simulation of runoff from the land surface to specified reaches. Groundwater evapotranspiration can be simulated with the SWR1 Process under model-based reaches using a linear, depth-dependent evapotranspiration function when simulated evaporation is less than specified evaporation rates. Aquifer-reach exchanges are a function of the difference between simulated stages and groundwater levels, and reach conductance. Conductance can be specified directly or calculated as a function of reach-bed sediment properties (leakance) and the maximum exchange perimeter, aquifer hydraulic conductivity, or a weighted combination of the leakance of reach-bed sediments and aquifer hydraulic conductivity. Reaches can be coupled to a single specified groundwater-model layer or dynamically coupled to multiple groundwater-model layers for reaches with defined geometries that intersect more than one groundwater-model layer.

Control structures are used to simulate the exchange of water between connected reaches. A variety of control structures are provided and include (1) an excess volume structure, (2) an uncontrolled-discharge structure, (3) a pump, (4) a rating-curve structure, (5) a culvert, (6) fixed and movable crest weirs, and (7) fixed and operable gated spillways. Multiple control structures can be implemented within individual reaches.

Solution of the nonlinear continuity equation at the reach-group scale (that is, a user defined collection of individual reaches) is achieved using Newton methods and either direct solution or Krylov sub-space methods (app. 3). A line-search algorithm also has been included to improve convergence and the quality of the Newton-step upgrade vector for highly nonlinear surface-water problems. Numerical stability can be improved through use of either linear- or sigmoid-depth scaling to increase reach-bed roughness and reduce reach outflows at small surface-water depths. Multiple SWR1 time steps can be simulated for each MODFLOW time step and an adaptive time-step algorithm, based on user-specified stage, flow, or SWR1 convergence constraints, has been implemented to decrease run times by minimizing the total number of SWR1 time steps needed during a simulation. Convergence of aquifer-reach exchanges between subsequent outer (Picard) MODFLOW time-step iterations can be enforced using a defined aquifer-reach criterion to ensure convergence of the aquifer-reach exchanges.

References Cited

Akan, A.O., and Yen, B.C., 1981, Diffusion-wave flood routing in channel networks: Journal of the Hydraulics Division: ASCE, v. 107, no. HY6, p. 719–732.

Alley, W.M., 2007, Another water-budget myth—The significance of recoverable groundwater in storage: Ground Water, v. 45, no. 3, p. 251.

Ansar, Matahel, and Chen, Zhiming, 2009, Generalized flow rating equations at prototype gated spillways: Journal of Hydraulic Engineering—ASCE, v. 135, no. 7, p. 602–608. (Also available at *http://dx.doi.org/10.1061/(ASCE)0733-9429(2009)135:7(602)*.)

Arnoldi, W.E., 1951, The principle of minimized iteration in the solution of the matrix eigenvalue problem: Quarterly of Applied Mathematics, v. 9, p. 17–29.

Barkau, R.L., 1996, UNET one-dimensional unsteady flow through a full network of open channels, User's Manual: U.S. Army Corps of Engineers CPD-66, version 3.1, 288 p.

Barrett, Richard, Berry, Michael, Chan, T.F., Demmel, James, Donato, June, Dongarra, Jack, Eijkhout, Victor, Pozo, Roldan, Romine, Charles, and Van Der Vorst, Henk, 1994, Templates for the solution of linear systems: Building blocks for iterative methods: Philadelphia, Pa., SIAM, 124 p.

Bedient, P.B., and Huber, W.C., 1988, Hydrology and floodplain analysis: Reading, Mass., Addison-Wesley, 650 p.

Brown, P.N., and Saad, Yousef, 1990, Hybrid Krylov methods for nonlinear systems of equations: SIAM Journal of Scientific and Statistical Computing, v. 11, no. 3, p. 450–481.

Burkardt, John, 2011, Fortran 90 reverse Cuthill-McKee ordering, accessed on October 5, 2011, at *http://people.sc.fsu.edu/~jburkardt/f_src/rcm/rcm.html*.

Chaudhry, M.H., 2008, Open-channel flow (2d ed.): New York, Springer, 523 p.

Chin, D.A., 1990, A method to estimate canal leakage to the Biscayne aquifer, Dade County, Fla.: U.S. Geological Survey Water-Resources Investigations Report 90-4135, 32 p.

Coon, W.F., 1998, Estimation of roughness coefficients for natural stream channels with vegetated banks: U.S. Geological Survey Water-Supply Paper 2441, 133 p.

Crout, P.D., 1941, A short method for evaluating determinants and solving systems of linear equations with real or complex coefficients: Transactions of the American Institute of Electrical Engineers, v. 60, p, 1235–1240.

Eisenstat, S.C., and Walker, H.F., 1996, Choosing the forcing terms in an inexact Newton method: SIAM Journal of Scientific Computing, v. 17, no. 1, p. 16–32.

Faunt, C.C., ed., 2009, Groundwater availability of the Central Valley aquifer, California: U.S. Geological Survey Professional Paper 1766, 225 p.

Feng Ke, and Molz, F.J, 1997, A 2-D, diffusion-based, wetland flow model: Journal of Hydrology, v. 196, p. 230–250.

French, R.H., 1985, Open-channel hydraulics: New York, McGraw Hill, 739 p.

Harbaugh, A.W., 2005, MODFLOW–2005, the U.S. Geological Survey modular ground-water model—The Ground-Water Flow Process: U.S. Geological Survey Techniques and Methods, book 6, chap. A16, variously paged.

Harbaugh, A.W., Banta, E.R., Hill, M.C., and McDonald, M.G., 2000, MODFLOW–2000, the U.S. Geological Survey modular ground-water model—User guide to modularization concepts and the Ground-Water Flow Process: U.S. Geological Survey Open-File Report 00-92, 121 p.

Harbaugh, A.W., and McDonald, M.G., 1996, User's documentation for MODFLOW–96, an update to the U.S. Geological Survey modular finite-difference ground-water flow model: U.S. Geological Survey Open-File Report 96-485, 56 p.

Hayami, Shōitirō, 1951, On the propagation of flood waves: Bulletin of the Disaster Prevention Research Institute, v. 1, no. 1, p. 1–16.

Henderson, F.M., 1966, Open channel flow: New York, MacMillan Publishing, 522 p.

Hill, M.C., 1990, Preconditioned conjugate-gradient 2 (PCG2), a computer program for solving ground-water flow equations: U.S. Geological Survey Water-Resources Investigations Report 90-4048, 43 p.

Hromadka T.V. II, McCuen, R.H., and Yen, C.C., 1987, Comparison of overland flow hydrograph models: Journal of Hydraulic Engineering—ASCE, v. 113, no. 11, p. 1422–1440.

Jobson, H.E. and Harbaugh, A.W., 1999, Modifications to the diffusion analogy surface-water flow model (DAFLOW) for coupling to the modular finite-difference ground-water flow model (MODFLOW): U.S. Geological Survey Open-File Report 99-217, 107 p.

Jones, J.E., and Woodward, C.S., 2001, Newton-Krylov-multigrid solvers for large-scale, highly heterogeneous, variably saturated flow problems: Advances in Water Resources, v. 24, p. 763–774.

Kelley, C.T., 1995, Iterative methods for linear and nonlinear equations—Frontiers in applied mathematics, v. 16: Philadelphia, Pa., SIAM, 166 p.

Kelley, C.T., 2003, Solving Nonlinear Equations with Newton's Method— Fundamental Algorithms for Numerical Calculations vol. 1: Philadelphia, Pa., SIAM, 104 p.

Knoll, D.A., and Keyes, D.E., 2004, Jacobian-free Newton-Krylov methods: a survey of approaches and applications: Journal of Computational Physics, v. 193, p. 357–397. (Also available at *http://dx.doi.org/10.1016/j.jcp.2003.08.010.*)

Krylov, A.N., 1931, On the numerical solution of the equation by which in technical questions frequencies of small oscillations of material systems are determined: Izvestiya Akademii Nauk SSSR, Otdelenie Matematicheskikh i Estestvennykh Nauk, v. 7, n. 4, p. 491–539.

Lal, A.M.W, 1998, Weighted implicit finite-volume model for overland flow: Journal of Hydraulic Engineering—ASCE, v. 124, no. 9, p. 941–950.

Lanczos, Cornelius, 1950, An iteration method for the solution of the eigenvalue problem of linear differential and integral operators: Journal of Research of the National Bureau of Standards, v. 45, p. 255–282.

Markstrom, S.L., Niswonger, R.G., Regan, R.S., Prudic, D.E., and Barlow, P.M., 2008, GSFLOW-Coupled Ground-water and Surface-water FLOW model based on the integration of the Precipitation-Runoff Modeling System (PRMS) and the Modular Ground-Water Flow Model (MODFLOW–2005): U.S. Geological Survey Techniques and Methods, book 6, chap. D1, 240 p.

McDonald, M.G., and Harbaugh, A.W., 1988, A modular three-dimensional finite-difference ground-water flow model: U.S. Geological Survey Techniques of Water-Resources Investigations, book 6, chap. A1, 586 p.

Merritt, L.M. and Konikow, L.F., 2000, Documentation of a computer program to simulate lake-aquifer interaction using the MODFLOW ground-water flow model and the MOC3D solute-transport model: U.S. Geological Survey Water-Resources Investigations Report 00-4167, 146 p.

Nemeth, M.S., and Solo-Gabriele, H.M., 2003, Evaluation of the use of reach transmissivity to quantify exchange between groundwater and surface water: Journal of Hydrology, v. 274, p. 145–159. (Also available at *http://dx.doi.org/10.1016/S0022-1694(02)00419-5.*)

Nemeth, M.S., Wilcox, W.M., and Solo-Gabriele, H.M., 2000, Evaluation of the use of reach transmissivity to quantify leakage beneath Levee 31N, Miami-Dade County, Florida: U.S. Geological Survey Water-Resources Investigations Report 00-4066, 80 p.

Newton, Isaac, 1671, Methodus fluxionum et serierum infinitorum: 678 p.

Niswonger, R.G. and Prudic, D.E., 2005, Documentation of the Streamflow-Routing (SFR2) Package to include unsaturated flow beneath streams—A modification to SFR1: U.S. Geological Survey Techniques and Methods, book 6, chap. A13, 47 p.

Niswonger, R.G., Prudic, D.E., and Regan, R.S., 2006, Documentation of the Unsaturated-Zone Flow (UZF1) Package for modeling unsaturated flow between the land surface and the water table with MODFLOW–2005: U.S. Geological Techniques and Methods, book 6, chap. A19, 62 p.

Otero, J.M., 1995, Computation of flow through water control structures: South Florida Water Management District Technical Publication no. 95-03 (WRE#328), 66 p.

Panday Sorab, and Huyakorn, P.S., 2004, A fully coupled physically-based spatially-distributed model for evaluating surface/subsurface flow: Advances in Water Resources, v. 27, p. 361–382.

Pernice, Michael, and Walker, H.F., 1998, NITSOL: A Newton iterative solver for nonlinear systems: SIAM Journal of Scientific Computing, v. 19, no. 1, p. 302–318.

Picard, Émile, 1890, Mémoire sur la théorie des équations aux dérivées partielles et la méthode des approximations successives, Journal de mathématiques pures et appliquées, series 4, v. 6, p. 145–210.

Ponce, V.M., 1990, Generalized diffusion wave equation with inertial effects: Water Resources Research, v. 26, no. 5, p. 1099–1101.

Ponce, V.M., 1991, The kinematic wave controversy: Journal of Hydraulic Research—ASCE, v. 117, no. 4, p. 511–525.

Ponce, V.M., and Simons, B.D., 1977, Shallow wave propagation in open channel flow: Journal of the Hydraulics Division—ASCE, v. 103, no. HY12, p. 1461–1476.

Press, W.H., Teukolsky, S.A., Vetterling, W.T., and Flannery, B.P., 1990, Numerical recipes in C: New York, Cambridge University Press, 994 p.

Press, W.H., Teukolsky, S.A., Vetterling, W.T., and Flannery, B.P., 1999, Numerical recipes in Fortran 90, vol. 2: New York, Cambridge University Press, 500 p.

Prudic, D.E., Konikow, L.F., and Banta, E.R., 2004, A new stream-flow routing (SFR1) package to simulate stream-aquifer interaction with MODFLOW–2000: U.S. Geological Survey Open-File Report 2004-1042, 95 p.

Reilly, T.E., Dennehy, K.F., Alley, W.M., and Cunningham, W.L., 2008, Ground-water availability in the United States: U.S. Geological Survey Circular 1323, 70 p.

Saad, Yousef, 2003, Iterative methods for sparse linear systems (2d ed): Philadelphia, Pa., SIAM, 528 p.

Saad, Yousef, 1994a, ILUT: A dual threshold incomplete LU factorization: Numerical Linear Algebra with Applications, v. 1, p. 387–402. (Also available at *http://dx.doi.org/10.1016/S0022-1694(02)00419-5.*)

Saad, Yousef, 1994b, SPARSKIT: a basic tool kit for sparse matrix computations (version 2): Technical Report, Computer Science Department, University of Minnesota, Minneapolis, Minn, accessed July 29, 2010 at *http://www-users.cs.umn. edu/~saad/software/SPARSKIT/*.

Saad, Yousef, and Schultz, M.H., 1986, GMRES: A generalized minimal residual algorithm for solving nonsymmetric linear systems: SIAM Journal of Scientific and Statistical Computing, v. 7, no. 3, p. 856–869.

Saint-Venant, B. de, 1843, Memoir sur un mode d'interpolation applicable a des impossible des equations aux derives partielles: Comptes Rendus, v. 17, p. 1108–1115.

Schaffranek, R.W., 2004, Simulation of surface-water integrated flow and transport in two dimensions: SWIFT2D user's manual: U.S. Geological Survey Techniques and Methods, book 6, chap. B1, 115 p.

Schaffranek, R.W., Baltzer, R.A., and Goldberg, D.E., 1981, A model for simulation of flow in singular and interconnected channels: U.S. Geological Survey Techniques of Water-Resources Investigations, book 7, chap. 3, 110 p.

Swain, E.D., Howie, Barbara, and Dixon, Joann, 1996, Description and field analysis of a coupled ground-water/surface-water flow model (MODFLOW/BRANCH) with modifications for structures and wetlands in Southern Dade County, Florida: U.S. Geological Survey Water-Resources Investigations Report 96-4118, 67 p.

Swain, E.D., and Wexler, E.J., 1996, A coupled surface-water and ground-water flow model (MODBRANCH) for simulation of stream-aquifer interaction: U.S. Geological Survey Techniques of Water-Resources Investigations, book 6, chap. A6, 125 p.

Trefethen, L.N., and Bau, David, 1997, Numerical linear algebra: Philadelphia, Pa., SIAM, 361 p.

U.S. Army Corps of Engineers, 1994, Engineering and design—Flood Runoff Analysis, EM 1110-2-1417, p. 9–1 – 9–24.

U.S. Army Corps of Engineers, 2008, HEC-RAS river analysis system—User's manual version 4.0: CDP-68, 733 p.

Zoppou, C., 1999, Reverse routing of flood hydrographs using level pool routing: Journal of Hydrologic Engineering, v. 4, no. 2, p. 184–188.

Appendixes

Appendix 1.   Notation Used in this Report

The symbol, first occurrence of each symbol, dimensions, and a description of the symbol are included.

Generic Units

[unitless]	Dimensionless
[1]	Unity—implies dimensionless
[0]	Zero—implies dimensionless
[L]	Length units
[M]	Mass units
[T]	Time units
[°]	Degrees

Special Notation

Symbol	Description
Ψ	Generic variable
$\dfrac{\partial \Psi}{\partial t}$ or $\dfrac{d\Psi}{dt}$	Time derivative of Ψ
$\dfrac{\partial \Psi}{\partial x}$ or $\dfrac{d\Psi}{dx}$	Space derivative of Ψ
$\Delta\Psi$	Discrete change in Ψ (for example, $\Delta\Psi = \Psi_1 - \Psi_2$)
$i = \overline{1, nn} = 1, 2, 3, 4, ..., nn$	Index i takes on all integer values between one and nn
$\displaystyle\sum_{i=1}^{nn} \Psi_i = \Psi_1 + \Psi_2 + \Psi_3 + ... + \Psi_{nn}$	Summation of Ψ
$\mathrm{sign}\left(\dfrac{\partial \Psi}{\partial x}\right)$	-1 if $\dfrac{\partial \Psi}{\partial x} < 0$ and +1 if $\dfrac{\partial \Psi}{\partial x} \geq 0$
$|\Psi|$	Absolute value of Ψ
$\min|\Psi_1, \Psi_2|$	Minimum of Ψ_1 and Ψ_2
$\mathbf{v}$	Vector
$\mathbf{M}$	Matrix
$\|\mathbf{v}\|_2$	ℓ^2 norm of vector $\mathbf{v}$
e^x	Exponential function

Roman Lowercase

Symbol	First occurrence	Dimensions	Description
a_{nm}	Eq. 3–10	[L^2/TL]	Off-diagonal element of Jacobian
b'	Eq. 48	[L]	Thickness of reach-bed sediments
c	Eq. 66	[L/T]	Wave celerity
c_M	Eq. 6	[Unitless]	Conversion factor for converting Manning's n roughness coefficient from $s/m^{1/3}$ to other length and/or time units
c_3	Eq. 38	[Unitless]	Submergence exponent
d	Eq. 2	[L]	Flow depth
d_o	Eq. 68	[L]	Reference flow depth
d_C	Eq. 35	[L]	Culvert diameter
d_{hC}	Eq. 33	[L]	Hydraulic depth of culvert
d_{min}	Fig. 11	[L]	Minimum flow depth above the structure invert elevation on the headwater or tailwater side of a control structure
d_{max}	Fig. 11	[L]	Maximum flow depth above the structure invert elevation on the headwater or tailwater side of a control structure
d_{VW}	Eq. 46	[L]	Flow depth of v-notch weir
f_h	Eq. 59	[L]	Adaptive time stepping stage change factor
f_{Inf}	Eq. 60	[L]	Adaptive time stepping inflow change factor
$f_{AQ_{max}}$	Eq. 64	[Unitless]	User-specified maximum allowed relative difference in SWR1 and MODFLOW aquifer-reach exchanges
f_n	Eq. 55	[L]	Linear- or sigmoid-depth scaling function
g	Eq. 2	[L/T^2]	Gravitational acceleration
h	Eq. 2	[L]	Surface-water stage
$\mathbf{h}$	Eq. 3–1	[Unitless]	State vector – simulated stage
$h_{initial}$	Eq. 53	[L]	Initial reach stage
h_G	Fig. 12	[L]	Structure gate elevation
h_{Gmax}	Eq. 44	[L]	Maximum control structure gate elevation
h_m	Eq. 53	[L]	Simulated stage in reach group m
h_{max}	Fig. 11	[L]	Maximum stage on headwater and tailwater side of a control structure
h_{min}	Fig. 11	[L]	Minimum stage on headwater and tailwater side of a control structure
h_{offset}	Eq. 53	[L]	Reach stage offset
h_S	Fig. 11	[L]	Structure invert elevation
is	Eq. 19	[Unitless]	Structure number in reach ii
ir_{max}	Eq. 3–13	[Unitless]	User-specified maximum number of Krylov iterations
k_{max}	Eq. 3–15	[Unitless]	User-specified maximum number of non-linear (outer) iterations
$klmax$	Eq. 47	[Unitless]	Maximum layer intersected by reach geometry
$klmin$	Eq. 47	[Unitless]	Minimum layer intersected by reach geometry
l_C	Eq. 29	[L]	Culvert length parallel to flow
l_{REACH}	Eq. 49	[L]	Length of reach
$l_{R \rightarrow node}$	Eq. 50	[L]	Horizontal distance from the center of the reach to the center of the grid cell
$nconn$	Eq. 17	[Unitless]	Number of reach connections in a reach group
nrg	App. 3	[Unitless]	Number of reach groups
nrg_r	App. 3	[Unitless]	Maximum bandwidth of $\mathbf{J}$ – equals 1 plus 2 times the maximum distance between the diagonal and connected reach groups in each linear equation (row)
n_C	Eq. 29	[$T/L^{1/3}$]	Culvert Manning's roughness coefficient

Symbol	First occurrence	Dimensions	Description
$\overline{n}$	Eq. 18	$[T/L^{1/3}]$	Distance weighted Manning's roughness coefficient
n_M	Eq. 6	$[T/L^{1/3}]$	Manning's roughness coefficient
nr	Eq. 64	[Unitless]	Number of reaches
ns	Eq. 19	[Unitless]	Number of structures in reach ii
q_{AQ}	Eq. 1	$[L^3/TL]$	Volumetric flow rate of aquifer-reach exchange per unit length
q_{EV}	Eq. 1	$[L^3/TL]$	Volumetric flow rate of evaporation per unit length
q_{LAT}	Eq. 1	$[L^3/TL]$	Volumetric flow rate of inflow and outflow from internal or external lateral sources per unit length
q_{PR}	Eq. 1	$[L^3/TL]$	Volumetric flow rate of rainfall per unit length
r_C	Fig. 39	$[T]$	Culvert radius
$\mathbf{r}_o$	Eq. 3–11	$[L^3/T]$	Initial residual at start of inner iteration
s	Eq. 12	$[L]$	is the relative spatial coordinate in the direction of maximum local stage slope
t	Eq. 1	$[T]$	Time
v	Eq. 66	$[L/T]$	Flow velocity
$\mathbf{v}$	Eq. 3–7	[Unitless]	Upgrade vector for simulated stage
v_o	Eq. 67	$[L/T]$	Reference mean flow velocity
v_{OPR}	Eq. 27	$[L]$ or $[L^3/T]$	Simulated stage or flow in the user-defined reach OPR
v_{SC}	Fig. 11	$[L]$ or $[L^3/T]$	Structure control stage or flow
w_p	Eq. 7	$[L]$	Wetted perimeter
x	Eq. 1	$[L]$	Spatial coordinate in the x-direction
y	Eq. 13	$[L]$	Spatial coordinate in the y-direction
$\mathbf{y}$	Eq. 3–8	$[L]$	Product of the upper triangular and upgrade vector – used in LU decomposition
z	Eq. 2	$[L]$	Reach bottom elevation
z_{max}	Eq. 33	$[L]$	Invert elevation of culvert on upstream side (h_{max} side) of the culvert

Roman Uppercase

Symbol	First occurrence	Dimensions	Description
A	Eq. 1	$[L^2]$	Cross-sectional area
$\overline{A}$	Eq. 18	$[L^2]$	Distance weighted cross-sectional area
A_C	Eq. 28	$[L^2]$	Culvert cross-sectional area
A_r	Eq. 48	$[L]$	Reach bottom surface area
C	Eq. 47	$[L^2/T]$	Aquifer-reach conductance
C_C	Eq. 28	[Unitless]	Culvert discharge coefficient
C_D	Eq. 37	[Unitless]	Weir discharge coefficient
C_F	Eq. 37	[Unitless]	Submergence factor
C_O	Eq. 42	[Unitless]	Orifice discharge coefficient
C_{RISE}	Eq. 31	$[L]$	Culvert rise (height)
C_{SPAN}	Eq. 32	$[L]$	Culvert span (width)
C_T	Eq. 42	[Unitless]	Gated spillway transition discharge coefficient
C_W	Eq. 39	$[L^{1/2}/T]$	Dimensional weir discharge coefficient
D_C	Eq. 28	$[L]$	Diameter of a circular culvert or rise of rectangular culvert
D_h	Eq. 67	$[L^2/T]$	Noninertial hydraulic diffusivity
$\tilde{D}_h$	Eq. 67	$[L2/T]$	Actual (inertial) hydraulic diffusivity

Symbol	First occurrence	Dimensions	Description
D_M	Eq. 10	[L/T]	Nonlinear diffusion term
D_{Mx}	Eq. 12	[L/T]	Nonlinear diffusion term in the x-direction
D_{My}	Eq. 13	[L/T]	Nonlinear diffusion term in the y-direction
D_{nn}	Eq. 3–10	[L²/TL]	Diagonal element of Jacobian
DMINDPTH	Eq. 55	[L]	User-defined minimum reach water depth
DUPDPTH	Eq. 55	[L]	Maximum reach water depth – defined from *DMINDPTH*
ETEXTD	Eq. 63	[L]	Cutoff depth or extinction depth for groundwater evapotranspiration
F	Eq. 3–1	[L³/T]	Residual form of the continuity equation
F'	Eq. 3–2	[L³/T]	Derivative of the residual form of the continuity equation
Fr	Eq. 66	[Unitless]	Froude number
G_O	Fig. 12	[L]	Structure gate opening
H	Eq. 47	[L]	Groundwater head
I	Eq. 3–12	[Unitless]	Identity matrix
Inf	Eq. 60	[L³/T]	Volumetric inflow rate
IA	Eq. 3–16	[Unitless]	Integer pointer to the first storage element for each reach group
IU	Eq. 3–16	[Unitless]	Integer pointer to the position of the first upper triangular (**U**) storage element for each reach group
J	Eq. 3–3	[L³/TL]	Jacobian – change in continuity equation residual with a small stage perturbation
J̃	Eq. 3–5	[L³/TL]	Approximate Jacobian - Finite-difference form of Jacobian
JA	Eq. 3–16	[Unitless]	Integer pointer to the columns for each reach group
K_b	Eq. 48	[L/T]	Hydraulic conductivity of reach-bed sediments
K'_{REACH}	Eq. 49	[1/T]	Leakance of reach bed sediments
$K_{H_{j,i,k}}$	Eq. 50	[L/T]	Horizontal hydraulic conductivity of groundwater model
$\mathbf{K}_r$	Eq. 3–11	[L³/T]	Krylov sub-space
L	Eq. 3–7	[L³/TL]	Lower triangular matrix of **J**
L_{OPR}	Fig. 11	[Unitless]	Logical operand for control structure; 0 = less than; 1 = greater than or equal to
M	Eq. 3–12	[L³/TL]	Preconditioned form of approximate Jacobian
M_a	Eq. 3	[L³/T²]	Momentum changes associated with local and convective acceleration terms
$M_{q_{AQ}}$	Eq. 2	[L³/T²]	Momentum change due to aquifer-reach exchanges
$M_{q_{LAT}}$	Eq. 2	[L³/T²]	Momentum change due to lateral inflows or outflows
M_0	Eq. 4	[Unitless]	Dimensionless momentum change calculated as the total momentum change divided by the product of gravitational acceleration and cross-sectional area
OPR	Fig. 11	[Unitless]	Reach used to control structure operation
Q	Eq. 1	[L³/T]	Streamflow or volumetric flow rate
Q_{AQ}	Eq. 15	[L³/T]	Volumetric flow rate of aquifer-reach exchange
Q_{BS}	Eq. 17	[L³/T]	Volumetric flow rate to external boundaries from surface-water control structures that are not connected to another reach
Q_{CS}	Eq. 17	[L³/T]	Volumetric flow rate required to balance the inflows and outflows for constant-stage reaches
Q_{error}	Eq. 65	[L³/T]	Volumetric surface-water budget error
Q_{EV}	Eq. 15	[L³/T]	Volumetric flow rate of evaporation
$Q_{EVactual}$	Eq. 63	[L³/T]	Simulated volumetric flow rate of reach evaporation
Q_{EXT}	Eq. 65	[L³/T]	Volumetric external flow rate programmatically applied using the `GWF2SWR7EX` subroutine

Symbol	First occurrence	Dimensions	Description
Q_{GWET}	Eq. 63	[L³/T]	Volumetric flow rate of evapotranspiration from groundwater
Q_{LAT}	Eq. 15	[L³/T]	Volumetric flow rate of inflow and outflow from internal or external lateral sources
Q_M	Eq. 6	[L³/T]	Volumetric flow rate between connected reaches
Q_{MAX}	Fig. 11	[L³/T]	Maximum volumetric flow rate for control structure
$Q_{M_{INFLOW}}$	Eq. 65	[L³/T]	Volumetric flow rate entering the reach group
$Q_{M_{OUTFLOW}}$	Eq. 65	[L³/T]	Volumetric flow rate leaving the reach group
Q_{MF}	Eq. 64	[L³/T]	Volumetric flow rate of aquifer-reach exchange calculated by MODFLOW
Q_{Mx}	Eq. 12	[L³/T]	Volumetric flow rate between connected reaches in the x-direction
Q_{My}	Eq. 13	[L³/T]	Volumetric flow rate between connected reaches in the y-direction
Q_{PR}	Eq. 15	[L³/T]	Volumetric flow rate of rainfall
Q_S	Eq. 19	[L³/T]	Volumetric structure flow rate
Q_{SL}	Eq. 65	[L³/T]	Volumetric specified lateral flow rate
Q_{UZF}	Eq. 65	[L³/T]	Volumetric dynamic inflow rate calculated by the MODFLOW UZF1 Package
R	Eq. 6	[L]	Hydraulic radius
R_C	Eq. 31	[L]	Hydraulic radius of culvert
$\bar{R}$	Eq. 18	[L]	Distance weighted hydraulic radius
RAI_{max}	Eq. 57	[L/T]	Maximum rainfall rate allowed using Δt_{max}
$\mathbf{RAI}_n$	Eq. 57	[L/T]	Rainfall rate applied to SWR1 reaches in current MODFLOW stress period
$Rate_G$	Eq. 44	[L/T]	Control structure gate operation rate
S_f	Eq. 2	[Unitless]	Reach friction slope
S_o	Eq. 68	[Unitless]	Reach bottom slope
S_{TYPE}	Fig. 11	[L]	Structure invert elevation
$\mathbf{U}$	Eq. 3–7	[L³/TL]	Upper triangular matrix of $\mathbf{J}$
V	Eq. 15	[L³]	Volume
$V_{v_{sc}}$	Eq. 24	[L³]	Available volume above the control elevation on the headwater side of the structure
W_S	Eq. 37	[L]	Weir length perpendicular to flow

Greek Lowercase

Symbol	First occurrence	Dimensions	Description
α	Eq. 67	[Unitless]	Factor that defines the relation of the Froude number and inertial terms
β	Eq. 2	[Unitless]	Momentum correction factor
ε_{mach}	Eq. 3–6	[Unitless]	Machine epsilon or unit roundoff
η	Eq. 3–13	[Unitless]	Nonlinear forcing function for Newton step for Krylov methods
γ	Eq. 3–15	[Unitless]	Dampening parameter for nonlinear forcing function for Newton step for Krylov methods
λ	Eq. 3–17	[Unitless]	Line search (backtracking) scaling factor
μ	Eq. 3–17	[Unitless]	User-specified dampening/acceleration factor
σ	Eq. 3–5	[L]	Stage perturbation
θ_C	Eq. 34	[°]	Angle between the center of the culvert and the top of the water surface in the culvert
θ_{VW}	Eq. 46	[°]	Angle of v-notch in v-notch weir
χ_p	Eq. 49	[L]	Maximum reach-aquifer exchange perimeter

Greek Uppercase

Symbol	First occurance	Dimensions	Description
Δh	Eq. 18	[L]	Stage difference
Δh_{max}	Eq. 59	[L]	User-specified, maximum allowed stage change criteria
Δh_S	Eq. 37	[L]	Controlling head difference over control structure (culvert, weir, spillway)
ΔI_{max}	Eq. 60	[L^3/T]	User-specified, maximum allowed inflow change criteria
Δt_{RAI}	Eq. 57	[T]	Maximum SWR1 time step for MODFLOW time step based on user-specified maximum reach rainfall rate
Δt	Eq. 17	[T]	SWR1 timestep
Δt_{max}	Eq. 57	[T]	User-specified, maximum SWR1 time step
Δt_{min}	Eq. 57	[T]	User-specified, minimum SWR1 time step
$\Delta t_{increment}$	Eq. 58	[Unitless]	User-specified, SWR1 time step adjustment factor
ΔV	Eq. 65	[L^3]	Volume change
Δx	Eq. 18	[L]	Distance between the center of two reach groups

Superscript

Symbol	First occurrence	Dimensions	Description
-1	Eq. 3–12	[Unitless]	Inverse of matrix
ir	Eq. 3–11	[Unitless]	Krylov iteration number
k	Eq. 3–2	[Unitless]	Nonlinear iteration index
T	Eq. 3–14	[Unitless]	Transpose of vector or matrix
t	Eq. 17	[T]	Previous time step
$t+\Delta t$	Eq. 17	[T]	Current time step
$+$	Eq. 58	[Unitless]	Variable estimate for current outer iteration

Subscript

Symbol	First occurrence	Dimensions	Description
i	Eq. 47	[Unitless]	MODFLOW row i
j	Eq. 47	[Unitless]	MODFLOW column j
k	Eq. 47	[Unitless]	MODFLOW layer k
ii	Fig. 11	[Unitless]	Current reach ii
jj	Fig. 11	[Unitless]	Connected reach jj
ii-jj	Fig. 11	[Unitless]	Connection between current reach ii and connected reach jj
n-m	Eq. 18	[Unitless]	Connection between current reach group n and connected reach group m
x	Eq. 12	[Unitless]	Variable value in the x-direction
y	Eq. 13	[Unitless]	Variable value in the y-direction

Appendix 2.   Data Input Instructions for the Surface-Water Routing (SWR1) Process

The use of the surface-water routing process (SWR1) is an advanced application of MODFLOW and it is assumed that users are familiar with the use of MODFLOW and the input files required for MODFLOW as documented in Harbaugh (2005), thus this appendix only describes input files required by SWR1.

MODFLOW Name File

Surface-water routing is activated by including a record in the MODFLOW Name file using the file type (Ftype) "SWR" to indicate that relevant calculations are to be made in the model and to specify the related input data file.

Surface-Water Routing Process Input Data

The SWR1 Process input file consists of input items numbered 0 through 14, each consisting of one or more records. These data are used to specify information about the locations, interconnections, flows, physical and hydraulic properties of surface-water features, as well as specifying certain output control options.

All input variables are read using free format, unless specifically indicated otherwise. In free format, variables are separated by one or more spaces, or by a comma and (optionally) one or more spaces. Thus, in free format, a blank field cannot be used to set a variable value to zero.

When directly inputting list data for datasets in the SWR1 input file, one line of a file is used for each reach for which data are specified. Each line explicitly includes the data values and any necessary indices. For example, data set 6 includes two types of data for each reach (a reach index and a boundary type flag), and each line of data includes both data types. The order of the entries in the list do not matter except for data sets 11b, 11c, and 13b. The direct approach of reading list data is implemented by a SWR1 utility subroutine (SSWRLSTRD), which provides a common mechanism for reading list data required for the SWR1 Process. Detailed input instructions for SSWRLSTRD are contained in the Input Instructions for SWR1 List Utility Subroutine section of this appendix.

When directly inputting two-dimensional real data, the form of input is to read the values one row at a time, starting with the first row. Within a row, columns are read across a line of a file starting with column one. If one line cannot hold all columns in a row, additional lines are used as required. Because the data are read in this prescribed row/column order, the row and column do not need to be explicitly indicated.

The direct approach for reading two-dimensional real data is implemented by the MODFLOW utility U2DREL subroutine, which provides a common mechanism for reading the data required for the SWR1 Process. This subroutine starts by reading a "control line," which is read from the data file for the process that requires the layer data. If all the cells have the same value, the value is specified on the control line, and reading separate values for each cell is not necessary. If the values vary among cells, lines containing the values for all cells are read either from the SWR1 input file, or from the file indicated in the control line. The format for reading the values also is specified in the control line. Thus, a great deal of flexibility is allowed regarding the organization of the input data for a simulation. Detailed input instructions for U2DREL are contained in the Input Instructions for Array Reading Utility Subroutines section of the MODFLOW-2005 documentation (Harbaugh 2005, p. 8-57 – 8-59).

Use of aquifer properties to define conductance values for surface-water features in the SWR1 Process relies on either the horizontal (HY) hydraulic conductivity in the Block-Centered Flow (BCF) Package or the horizontal (HK) hydraulic conductivity data in the Layer Property Flow (LPF) or the Hydrogeologic-Unit Flow (HUF) Packages. When using the BCF Package, the right digit of LTYPE (LAYCON) must be equal to one (1) or three (3), or the model will yield an error and stop execution.

External time series files can be used to define rainfall (RAIN), evaporation (EVAP), lateral flows (QLATFLOW), control-elevation criterion (STRCRIT), and specified discharge or specified gate opening data (STRVAL) for operable control structures for select reaches and structures. SWR1 time-step lengths can also be specified using an external time series file. Including the USE_TABFILES keyword in Item 1b and adding a tabular data section in the SWR1 input file results in use of a SWR1 utility subroutine (SSWR_RDTABDATA) to process time series data. The SSWR_RDTABDATA subroutine provides a common mechanism for reading optional time series data in the SWR1 Process from ASCII and binary (UNFORMATTED) files. Multiple external time series files can be specified but must be included in the NAM file. Detailed input instructions for SSWR_RDTAB-DATA are contained in the Input Instructions for SWR1 Time Series Utility Subroutine section of this appendix.

The SWR1 Process does not support use of parameters, but functionality has been included to define surface-water geometry entries that can be distributed to multiple reaches. Functionality also has been added to the SWR1 Process to permit inclusion of comments (denoted with a # in column 1) and blank lines between records for Items 0 through 14:

FOR EACH SIMULATION

0.   [#Text]

Item 0 is optional—# must be in column 1. Item 0 can be repeated multiple times.

1a. NREACHES ISWRONLY ISWRCBC ISWRPRGF ISWRPSTG ISWRPQAQ ISWRPQM ISWRPSTR ISWRPFRN [Option]

If SWROPTIONS option specified in Item 1a:

1b. CSWROPT [IOPTUNIT]

Item 1b is read only if a non-zero value is specified for optional variable ISWROPT is specified in Item 1a. Item 1b can include multiple SWR1 options, on separate lines, but must be terminated using the keyword "END".

2. DLENCONV TIMECONV RTINI RTMIN RTMAX RTPRN RTMULT NTMULT DMINGRAD DMNDEPTH DMAXRAI DMAXSTG DMAXINF

3. ISOLVER NOUTER NINNER IBT TOLS TOLR TOLA DAMPSS DAMPTR IPRSWR MUTSWR [IPC] [NLEVELS] [DROPTOL] [IBTPRT]

4a. IRCH4A IROUTETYPE IRGNUM KRCH IRCH JRCH RLEN

4b. IRCH4B NCONN ICONN(1)...ICONN(NCONN)

If USE_TABFILES optional keyword is specified in Item 1a:

4c. NTABS

If USE_TABFILES optional keyword is specified in Item 1a and NTABS > 0:

4d. CTABTYPE ITABUNIT CINTP [CTABRCH] [ITABRCH(1)...ITABRCH(NTABRCH)]

Item 4a and 4b is read by subroutine SSWRLSTRD and must be repeated NREACHES times.

Item 4c and 4d are read only if the USE_TABFILES optional keyword is specified in Item 1b. Item 4d must be repeated NTABS times.

The stream network is assumed to remain fixed in space over the duration of a simulation. However, the active part of the stream network can be made to vary over time by making selected reaches inactive for selected stress periods. Inactive reaches would be specified by setting the reach ISWRBND in Item 6 to zero for the specific stress periods when they are known to be inactive or dry.

If the finite-difference cell corresponding to a surface-water reach is inactive, no aquifer-reach exchanges are allowed.

FOR EACH STRESS PERIOD

5. ITMP IRDBND IRDRAI IRDEVP IRDLIN IRDGEO IRDSTR IRDSTG IPTFLG [IRDAUX]

If ITMP > 0, read Items 6, 7, 8, 9, 10, 11, 12, and/or 13 based on values specified in Item 5:

If IRDBND > 0:

6. IBNDRCH ISWRBND

Item 6 is read by subroutine SSWRLSTRD and must be repeated IRDBND times. IRDBND must be equal to NREACHES on the first stress period. If IRDBND = 0 reach boundary data from the previous stress period will be reused.

If |IRDRAI| > 0:

7a. IRAIRCH RAIN

7b. [RAIN2D (NCOL, NROW)] -- U2DREL

If IRDRAI = 0 reach rainfall data from the previous stress period will be reused. Item 7a is read by subroutine SSWRLSTRD, must be repeated IRDRAI times, and is only specified if IRDRAI > 0. Item 7b is read by subroutine U2DREL and is only specified if IRDRAI < 0.

If |IRDEVP| > 0:

8a. IEVPRCH EVAP

8b. [EVAP2D (NCOL, NROW)] -- U2DREL

If IRDEVP = 0 reach evaporation data from the previous stress period will be reused. Item 8a is read by subroutine SSWRLSTRD, must be repeated IRDEVP times, and is only specified if IRDEVP > 0. Item 8b is read by subroutine U2DREL and is only specified if IRDEVP < 0.

If |IRDLIN| > 0:

9a. ILINRCH QLATFLOW

9b. [QLATFLOW2D (NCOL, NROW)] -- U2DREL

If IRDLIN = 0 reach lateral flow data from the previous stress period will be reused. Item 9a is read by subroutine SSWRLSTRD, must be repeated IRDLIN times, and is only specified if IRDLIN > 0. Item 9b is read by subroutine U2DREL and is only specified if IRDLIN < 0.

If IRDGEO > 0:

10. IGMODRCH IGEONUMR GZSHIFT

Item 10 is read by subroutine SSWRLSTRD and must be repeated IRDGEO times. IRDGEO must be equal to NREACHES on the first stress period. If IRDGEO < 1 geometry data from the previous stress period will be reused.

If IRDGEO > 0:

11a.   IGEONUM IGEOTYPE IGCNDOP GMANNING [NGEOPTS] [GWIDTH] [GBELEV] [GSSLOPE] [GCND] [GLK] [GCNDLN] [GETEXTD]

If **IGEOTYPE** = 3 repeat NGEOPTS times:

11b.   XB(1) ELEVB(1)

XB(2) ELEVB(2)

.

.

.

XB(NGEOPTS) ELEVB(NGEOPTS)

If **GEOTYPE** = 4 repeat NGEOPTS times:

11c.   ELEV(1) VOL(1) WETPER(1) SAREA(1) XAREA(1)

ELEV(2) VOL(2) WETPER(2) SAREA(2) XAREA(2)

.

.

.

ELEV(NGEOPTS) VOL(NGEOPTS) WETPER(NGEOPTS) SAREA(NGEOPTS) XAREA(NGEOPTS)

If IRDGEO < 1 geometry data from the previous stress period will be reused. Item 11a to 11c must be repeated for each unique IGEONUM defined in Item 10. Item 11a will contain 6 to 10 variables; depending on the values of IGEOTYPE and IGCNDOP. IGEOTYPE defines the geometry type and IGCNDOP defines how the surface-water/groundwater conductance will be calculated.

Item 11b and 11c may include no input when all are defined by data in Item 11a or they may include as many as 5 variables depending if IGEOTYPE = 3 or 4. Item 11b and 11c are read by subroutine SSWRLSTRD and are repeated NGEOPTS times for each Item 11a entry (IRDGEO data) where IGEOTYPE = 3 or 4, respectively.

If IRDSTR > 0:

12. ISMODRCH NSTRUCT

   Item 12 is read by subroutine SSWRLSTRD and must be repeated IRDSTR times. If IRDSTR < 1, structure data from the previous stress period will be reused.

If IRDSTR > 0:

13a.   ISTRRCH ISTRNUM ISTRCONN ISTRTYPE [NSTRPTS] [STRCD] [STRCD2] [STRCD3]
[STRINV] [STRINV2] [STRWID] [STRWID2] [STRLEN] [STRMAN] [STRVAL] [ISTRDIR]

13b.   CSTROTYP ISTRORCH [ISTROQCON] CSTROLO CSTRCRIT [STRCRITC] [STRRT] STRMAX
[CSTRVAL]

13c.   STRELEV(1) STRQ(1)

STRELEV(2) STRQ(2)

.

.

.

STRELEV(NSTRPTS) STRQ(NSTRPTS)

   If IRDSTR < 1 structure data from the previous stress period will be reused. Items 13a, 13b, and 13c must be repeated the sum of NSTRUCT defined in Item 12. Item 13 will contain 5 to 12 variables depending on the values of ISTRTYPE, which defines the data requirements for the specific structure type. Item 13b may include no input when all are defined by data in Item 13a or it may include 5 to 9 variables. Item 13c may include no input when all are defined by data in Item 13a or they may include 2 variables depending if ISTRTYPE = 4. Item 13c is read by subroutine SSWRLSTRD and is repeated NSTRPTS times for each Item 13a entry (ISTRRCH-ISTRNUM combination) where ISTRTYPE = 4.

If |IRDSTG| > 0:

14a.   IRCHSTG STAGE

14b.   [STAGE2D (NCOL, NROW)] -- U2DREL

   If IRDSTG = 0 stage data from the previous stress period will be reused for constant-stage reaches. Item 14a is read by subroutine SSWRLSTRD and must be repeated IRDSTG times and is only specified if IRDSTG > 0. Item 14b is read by subroutine U2DREL and is only specified if IRDSTG < 0.

If IRDAUX > 0 and auxiliary variables specified in item 1a:

15. IRCHAUX [xyz]

   If IRDAUX = 0 auxiliary data from the previous stress period will be reused. Item 15 is read by subroutine SSWRL-STRD and must be repeated IRDAUX times and is only specified if IRDAUX > 0.

Explanation of Variables Read by the SWR1 Process

ITEM 0 VARIABLES:

Text—is a character variable (up to 199 characters) that starts in column 2. Any characters can be included in Text. The # character must be in column 1. Text is written to the Listing File.

ITEM 1a VARIABLES – Problem dimensions, unit conversions, and output controls:

NREACHES—an integer value equal to the number of surface-water reaches (features) that are active during the simulation. The value of NREACHES also represents the number of lines of data to be included in Item 4 and the number of times Item 4 must be repeated.

ISWRONLY—an integer value used as a flag for writing aquifer-reach leakage values. If ISWRONLY > 0, the MODFLOW boundary (IBOUND) array defined in the Basic (BAS) Package of the groundwater flow process is set to zero making all groundwater model cells inactive (see Harbaugh, 2005, p. 4-1 – 4-2) and disabling SWR1 aquifer-reach exchanges. If ISWRONLY ≤ 0, the groundwater process uses the boundary array defined in the BAS Package.

ISWRCBC—an integer value used as a flag for writing aquifer-reach leakage values and evapotranspiration values beneath model-based reaches. If ISWRCBC > 0, it is the unit number to which unformatted leakage between each reach and corresponding finite-difference cell will be saved to a file whenever the cell-by-cell budget has been specified in Output Control Option (see Harbaugh, 2005, p. 8-17—8-21). If ISWRCBC $= 0$, leakage values will not be printed or saved. If ISWRCBC < 0, aquifer-reach leakage values and evapotranspiration values beneath model-based reaches will be printed in the main listing file whenever a cell-by-cell budget has been specified in Output Control.

ISWRPRGF—an integer value used as a flag for writing to a separate formatted or unformatted binary file the stage and all information about inflows and outflows from each reach group. If ISWRPRGF > 0, then ISWRPRGF also represents the unit number to which all information for each reach group will be saved to a separate file using a comma-separated-value format. If ISWRPRGF < 0, |ISWRPRGF| is the unit number to which the stage and all information on inflows and outflows for each reach group will be saved to a binary file. Data for reach groups is saved to the |ISWRPRGF| unit number with a frequency equal to RTPRN or at the SWR1 time step corresponding to the end of each MODFLOW time step if RTPRN $= 0.0$.

ISWRPSTG—an integer value used as a flag for writing to a separate formatted or unformatted binary file the stage for each reach. If ISWRPSTG > 0, then ISWRPCFL also represents the unit number to which all reach stage data will be saved to a separate file using a comma-spaced-value format. If ISWRPSTG < 0, |ISWRPSTG| is the unit number to which the stage for each reach will be saved to a binary file. Data for reach stage is saved to the |ISWRPCFL| unit number with a frequency equal to RTPRN or at the SWR1 time step corresponding to the end of each MODFLOW time step if RTPRN $= 0.0$.

ISWRPQAQ—an integer value used as a flag for writing to a separate formatted or unformatted binary file the aquifer-reach exchanges from each reach; the stream stage, depth, wetted perimeter, and streambed conductance; and the head difference and gradient from the reach to the groundwater. If ISWRPQAQ > 0, then ISWRPQAQ also represents the unit number to which all reach aquifer-reach data will be saved to a separate file using a comma-spaced-value format. If ISWRPQAQ < 0, |ISWRPQAQ| is the unit number to which the aquifer-reach and related variables for each reach will be saved to a binary file. Aquifer-reach exchange data for each reach is saved to the |ISWRPQAQ| unit number with a frequency equal to RTPRN or at the SWR1 time step corresponding to the end of each MODFLOW time step if RTPRN $= 0.0$.

ISWRPQM—an integer value used as a flag for writing to a separate formatted or unformatted binary file the lateral flow and velocity for each reach connection. Non-zero velocities are only reported for reach connections without surface-water control structures. If ISWRPQM > 0, then ISWRPQM also represents the unit number to which all reach lateral flow and velocity data will be saved to a separate file using a comma-spaced-value format. If ISWRPQM < 0, |ISWRPQM| is the unit number to which the lateral flow and velocity data for each reach connection will be saved to a binary file. Lateral-flow data for each reach connection is saved to the |ISWRPQM| unit number with a frequency equal to RTPRN or at the SWR1 time step corresponding to the end of each MODFLOW time step if RTPRN $= 0.0$.

ISWRPSTR—an integer value used as a flag for writing to a separate formatted or unformatted binary file the structure flow, upstream stage, downstream stage, gate elevation, and gate opening data for each structure. If ISWRPSTR > 0, then ISWRPSTR also represents the unit number to which all structure data will be saved to a separate file using a comma-spaced-value format. If ISWRPSTR < 0, |ISWRPSTR| is the unit number to which the structure data will be saved to a binary file. Structure data are saved to the |ISWRPSTR| unit number with a frequency equal to RTPRN or at the SWR1 time step corresponding to the end of each MODFLOW time step if RTPRN = 0.0.

ISWRPFRN—an integer value used as a flag for writing summary information on the maximum Froude number for reaches simulated by the diffusive-wave approximation for each SWR1 time step to the MODFLOW LST file. A summary of the maximum Froude number for all SWR1 time steps is also summarized in the MODFLOW LST file at the end of the last MODFLOW time step in the last stress period.

Option—is an optional list of character values.

"AUXILIARY abc" or "AUX abc" —defines an auxiliary variable, named "abc", will be read for each SWR1 reach I item 15. Up to 20 variables can be specified, each of which must be preceeded by "AUXILIARY" or "AUX." These variables currently are not used by the SWR1 process, but they will be available for use by other processes.

"SWROPTIONS"—indicates optional SWR1 control variables will be specified in item 1b.

ITEM 1b VARIABLES – Optional control parameters:

CSWROPT—string value that represents a keyword for an optional SWR1 control variable. Available optional keywords include:

PRINT_SWR_TO_SCREEN, Iteration data for each SWR1 time step is reported to the screen.

SAVE_SWRDT, The SWR1 time step length for each SWR1 time step is written to an ASCII or binary output file.

SAVE_AVERAGE_RESULTS, Average simulated results for each SWR1 output time (RTPRN) will be written to ASCII and binary output files.

USE_TABFILES, Tabular data are specified in item 4c and 4d.

USE_NONCONVERGENCE_CONTINUE, The simulation will continue even if convergence is not achieved in the MODFLOW groundwater flow process and/or the SWR1 Process (non-convergence continuation option).

USE_INEXACT_NEWTON, An inexact Newton method will be used. See appendix 3 for further details on the inexact Newton method implement with the iterative Krylov solvers available with SWR1.

USE_STEADYSTATE_STORAGE, The pseudo-transient continuation approach is used for steady-state stress periods.

USE_LAGGED_OPR_DATA, The current estimates of reach stages and flows are used to surface water control structure operations.

USE_LINEAR_DEPTH_SCALING, Linear-depth scaling is used.

USE_DIAGONAL_SCALING, The Jacobian is scaled using symmetric diagonal scaling.

USE_L2NORM_SCALING, The Jacobian is scaled using symmetric ℓ^2-norm scaling.

USE_RCMREORDERING, Active reach groups are reordered using reverse Cuthill-Mckee reordering.

USE_RCMREORDERING_IF_IMPROVEMENT, Active reach groups are only reordered if reverse Cuthill-Mckee reordering reduces the Jacobian bandwidth.

END, Keyword for terminating item 1b. This keyword is required.

IOPTUNIT—as integer that represents the unit number to which the optional SWR1 output will be saved. If IOPTUNIT > 0, data will be saved to an ASCII file with unit number IOPTUNIT using a comma-spaced-value format. If IOPTUNIT < 0, |IOPTUNIT| is the unit number of the binary file to which the optional output data will be saved. IOPTUNIT is currently only specified for the CSWROPT keyword SAVE_SWRDT.

The order of keywords specified in item 1b is not important with the exception of the END keyword which must be the last entry in item 1b. Unrecognized keywords are ignored.

ITEM 2 VARIABLES – Solution controls:

DLENCONV— a real value (or conversion factor) used in converting flow terms that use gravitational acceleration to consistent length units. DLENCONV is used to calculate correct structure flows for fixed and operable surface-water structures defined in Item 8 (ISTRTYPE = 5, 6, 7, 8, and 9). DLENCON is also used to calculate correct discharges from active reaches using the diffusive-wave approximation (IROUTETYPE = 3). If fixed and operable control structures are not simulated and the diffusive-wave approximation is not used in a simulation (IROUTETYPE = 1 or 2), then a value of 1.0 can be entered. DLENCON should be set to 3.28081, 1.0, and 100.0 when using length units (LENUNI) of feet, meters, or centimeters in the simulation, respectively (see Harbaugh 2005, p. 3-7).

TIMECONV— a real value (or conversion factor) used in converting flow terms that use gravitational acceleration to consistent time units. DLENCONV is used to calculate correct structure flows for fixed and operable surface water structures defined in Item 8 (ISTRTYPE = 5, 6, 7, 8, and 9). TIMECONV is also used to calculate correct discharges from active reaches using the diffusive-wave approximation (IROUTETYPE = 3). If fixed and operable control structures are not simulated and the diffusive-wave approximation is not used in a simulation (IROUTETYPE = 1 or 2), then a value of 1.0 can be entered. TIMECONV should be set to 1.0, 60.0, 3,600.0, 86,400.0, and 31,557,600.0 when using time units (ITMUNI) of seconds, minutes, hours, days, or years in the simulation, respectively (see Harbaugh 2005, p. 3-7).

RTINI—a real value that defines the initial SWR1 time-step length to use in the first MODFLOW time step of the first stress period. RTINI should be ≥ RTMIN and ≤ RTMAX. Value is in units of time and must be consistent with the defined MODFLOW time unit.

RTMIN—a real value that defines the minimum SWR1 time-step length to use. RTMIN should be > 0.0. Typical values of RTMIN are between 1 and 30 seconds but is dependent on the magnitude of streamflow. Value is in units of time and must be consistent with the defined MODFLOW time unit.

RTMAX—a real value that defines the maximum SWR1 time-step length to use. RTMAX should be ≥ RTMIN and ≤ the minimum MODFLOW time-step length. Adaptive time stepping is disabled if RTMAX = RTMIN. Typical values of RTMAX are between 1 and 24 hours but are dependent on the magnitude of streamflow. Value is in units of time and must be consistent with the defined MODFLOW time unit.

RTPRN—a real value that defines the frequency of SWR1 results output to ASCII or binary result files (ISWRPCFL, ISWRPSTG, ISWRPQAQ, and/or ISWRPQM). Results are saved to SWR1 output files for SWR1 time steps corresponding to the end of each MODFLOW time step if RTPRN = 0.0. If RTPRN > 0.0, RTPRN should be ≥ RTMIN. Value is in units of time and must be consistent with the defined MODFLOW time unit.

RTMULT—a real value that defines the multiplier used to increase the SWR1 time-step length. Adaptive time stepping is disabled if RTMULT = 1.0. Typical values of RTMAX are between 1.0 and 1.2. RTMULT should be ≥ 1.

NTMULT—an integer value that defines the frequency of SWR1 time-step length increases. The SWR1 time-step length is increased only when SWR1 solver convergence is achieved, the SWR1 time-step length is less than the maximum time-step length for the stress period defined by DMAXRAI, the maximum stage change in any reach are less than DMAXSTG, and the maximum inflow to any reach is less than DMAXINF on NTMULT consecutive SWR1 time steps within a single MODFLOW time step. NTMULT values ranging from 2 to 10 have been found to work reasonably well in the test cases evaluated. NTMULT should be > 1.

DMINGRAD— a real value used to define the water-surface gradient that must be exceeded for surface-water flow to be simulated in reach groups or structures using the diffusive-wave approximation (IROUTETYPE = 3). DMINGRAD can be set to any real number if all reach groups in a simulations use the reservoir-routing approximation (IROUTETYPE = 1 or 2). A non-zero value should only be used in cases where small gradients are causing numerical instabilities or with SWR1 datasets with a large number of reach groups to reduce numerical overhead; in these cases DMINGRAD values ranging between 1.0×10^{-9} and 1.0×10^{-12} have been found to be reasonable.

DMINDPTH— a real value used to define the minimum reach depth where outflows (evaporation, aquifer-reach exchanges, and diffusive wave flow) are permitted. Sigmoid-depth scaling is the default depth scaling approach used. Scaling is not applied if DMINDPTH = 0.0. Value is in units of length and must be consistent with the MODFLOW length unit. A value of 1×10^{-3} m has been found to work well in the test simulations evaluated.

DMAXRAI— a real value used to define the maximum rainfall rate for SWR1 reaches in the current MODFLOW stress period. The SWR1 time step is reduced using equation 57 if the rainfall in any reach exceeds DMAXRAI. Rainfall is not used to adjust SWR1 time steps if DMAXRAI $\leq$ 0.0. DMAXRAI is only specified for SWR1 simulations with RTMIN < RTMAX and TMULT > 1.0. Value is in units of length per time and must be consistent with MODFLOW length and time units.

DMAXSTG— a real value used to define the maximum stage change permitted in a SWR1 time step. The SWR1 time step is reduced using equation 59 if the stage change in any reach exceeds DMAXSTG. Simulated stage changes are not used to adjust SWR1 time steps if DMAXSTG $\leq$ 0.0. DMAXSTG is only specified for SWR1 simulations with RTMIN < RTMAX and TMULT > 1.0. Value is in units of length and must be consistent with MODFLOW length unit.

DMAXINF— a real value used to define the maximum net inflow change permitted in a SWR1 time step. The SWR1 time step is reduced using equation 60 if the net inflow change in any reach exceeds DMAXINF. Simulated net inflow changes are not used to adjust SWR1 time steps if DMAXINF $\leq$ 0.0. DMAXINF is only specified for SWR1 simulations with RTMIN < RTMAX and TMULT > 1.0. Value is in units of cubic length per time (L^3/T) and must be consistent with MODFLOW length and time units.

ITEM 3 VARIABLES – Solver parameters:

ISOLVER—an integer value that defines the solver used by the SWR1 Process to solve the defined surface-water equations for each reach group.

ISOLVER = 1, LU decomposition using Crout's algorithm (Crout, 1941) based on Press and others (1990, 1999).

ISOLVER = 2, Biconjugate gradient stabilized method (bi-CGSTAB) based on Barrett and others (1994), and Kelley (1995).

ISOLVER = 3, GMRES method based on Brown and Saad (1990), Jones and Woodward (2001), and Saad and Schultz (1986).

NOUTER—an integer value that defines the maximum number of outer iterations used to resolve non-linearities for all solvers. NOUTER must be > 0 and is generally less than 100.

NINNER—an integer value that defines the maximum number of inner iterations used within the preconditioned biconjugate gradient solver (SOLVER = 2) and GMRES solver (SOLVER = 3). NINNER also defines the maximum number of iterations to perform before restarting GMRES (SOLVER = 3) and reduced storage requirements for the method. Typical values for the biconjugate gradient stabilized method range from 10 to 30. Typical values for the GMRES method range from 5 to 30. NINNER is not used for ISOLVER = 1 and can be set to any positive value greater than 0.

IBT—an integer value that defines whether line search is enabled (IBT > 1) or disabled (IBT $\leq$ 1). IBT also defines the maximum number of line search iterations to perform (if IBT > 1) at the end of each outer iteration to minimize the ℓ^2-norm using Brent's method (Press and others, 1990). Line search should be disabled (IBT $\leq$ 1) if DAMPSS and/or DAMPTR values other than 1.0 are specified.

TOLS—a real value that defines the tolerance level for reach group stages used to evaluate convergence for stage within the selected numerical solver and to trigger increases and decreases in the SWR1 time step when adaptive time stepping is enabled. Value is in units of length and must be consistent with the MODFLOW length unit. Usually a value of 1.0×10^{-9} is sufficient when units of feet or meters are used in the model.

TOLR—a real value that defines the tolerance level for reach group flow residuals to evaluate convergence for flow within the selected numerical solver and to trigger increases and decreases in the SWR1 time step when adaptive time stepping is enabled. Value is in units of length cubed per time, and must be consistent with MODFLOW length and time units. Usually a value of 1.0×10^{-1} is sufficient for the flow-residual criteria when meters and seconds are the defined MODFLOW length and time.

TOLA—a real value that defines the tolerance level for relative differences in surface-water/groundwater exchanges between SWR1 and MODFLOW at the end of each MODFLOW outer iteration (Picard iteration). Convergence of surface-water/groundwater exchanges between successive MODFLOW outer iterations is disabled if TOLA is less than or equal to zero. Disabling the tolerance level for surface-water/groundwater exchanges may be appropriate for models in which groundwater levels do not change noticeably from one MODFLOW time step to another; the validity of this assumption can be evaluated by looking at the differences in SWR1 and MODFLOW calculated aquifer-reach exchanges printed in the MODFLOW listing file. Use of TOLA may improve results in cases where use of lagged heads in SWR1 formulate subroutines (GWF2SWR1FM) allow convergence of MODFLOW's iterative solvers on the first outer iteration but will increase run-times. Value is relative and unitless. Usually a value of 0.01 is sufficient.

DAMPSS—a real value that defines the steady-state dampening factor that is applied for each inner iteration of the selected linear solver. Acceleration can be applied to the inner iteration of the selected linear solver by specifying DAMPSS values greater than 1.0. DAMPSS values between 0.1 and 0.5 have been found to work well for steady-state stress periods. Dampening/acceleration is not applied if a DAMPSS value of 0.0 is specified (DAMPSS is reset to 1.0). DAMPSS should be set to 1.0 if IBT is greater than 1.

DAMPTR—a real value that defines the transient dampening factor that is applied for each inner iteration of the selected linear solver. Acceleration can be applied to the inner iteration of the selected linear solver by specifying DAMPTR values greater than 1.0. Dampening/acceleration is not applied if a DAMPTR value less than or equal to 0.0 is specified (DAMPTR is reset to 1.0). DAMPTR should be set to 1.0 if IBT is greater than 1.

IPRSWR—an integer value used as the printout interval for convergence information for the SWR1 Process. SWR1 convergence information is written to the MODFLOW listing file every IPRSWR MODFLOW time steps. If IPRSWR is less than 1 it is changed to 999. This printout also occurs at the end of each MODFLOW stress period regardless of the value of IPRSWR.

MUTSWR—an integer value used as a flag that controls printing of convergence information for the SWR1 Process:
MUTSWR = 0, is for printing tables of the final maximum absolute residual for each SWR1 time step.
MUTSWR = 1, is for printing only the total number of iterations in each SWR1 time step.
MUTSWR = 2, is for no printing.
MUTSWR = 3, is for printing only if convergence fails for at least one SWR1 time step in a MODFLOW time step.

IPC—an integer value that defines the preconditioner with the iterative solver used by the SWR1 Process to solve the defined surface-water equations for each reach group. IPC is only defined if ISOLVER > 1.

IPC = 0, No preconditioning.

IPC = 1, Jacobi preconditioning.

IPC = 2, ILU(0) preconditioning.

IPC = 3, MILU(0) preconditioning.

IPC = 4, ILUT preconditioning.

NLEVELS—an integer value that defines the maximum number of levels to use for each row of **U** and each row of **L**. NLEVELS values between 5 and 10 have been found to work well for the test problems evaluated. NLEVELS is only specified if IPC = 4.

DROPTOL—a real value that defines the threshold for dropping small terms in the factorization. A DROPTOL value of 1×10^{-3} has been found to work well for the test problems evaluated. DROPTOL is only specified if IPC = 4.

IBTPRT—an integer value used as a flag for writing a summary of line-search information to the MODFLOW listing file for each SWR1 outer iteration when line search is attempted. Line-search information summarized includes the SWR1 outer iteration number, number of line search iterations, initial ℓ^2-norm of the residual, final ℓ^2-norm of the residual, final relaxation parameter value (λ), and whether the estimated stage was updated at the end of the line search routine. If ISWRSOL > 0, then summary line-search data are written to the MODFLOW listing file every IBTPRT SWR1 outer iteration when line search is attempted.

ITEM 4a VARIABLES – Time-invariant reach data:

IRCH4A— an integer value equal to the number of the reach. IRCH4A must be greater than zero and less than or equal to NREACHES. IRCH4A data can be specified in any order but each entry must be assigned a unique number and be in the range of 1 to NREACHES.

IROUTETYPE—an integer value that defines the surface-water equation used in each reach.

> IROUTETYPE = 1, Level-pool reservoir routing.
>
> IROUTETYPE = 2, Tilted-pool reservoir routing.
>
> IROUTETYPE = 3, Diffusive-wave approximation.

IRGNUM— an integer value for defining a reach group number for the reach. IRGNUM can be any integer value.

KRCH—an integer value equal to the layer number of the cell containing the reach. If KRCH < 1, the reach can span multiple layers that are defined using geometric data specified in Item 8 for the reach. KRCH must be set to 1 for reaches that cover an entire finite-difference cell (IGEOTYPE = 5; see Item 11a).

IRCH—an integer value equal to the row number of the cell containing the reach.

JRCH—an integer value equal to the column number of the cell containing the reach.

RLEN—a real number equal to the length of reach within this finite-difference cell. The length of a reach can exceed the finite-difference cell dimensions because of the meandering nature of many one-dimensional surface-water features. The length is used to calculate the reach conductance for this reach for several leakage options. RLEN can be defined as any value for reaches defined as IGEOTYPE = 4 and 5 (see the description of Item 11 variables).

ITEM 4b VARIABLES – Time-invariant reach connectivity data:

IRCH4B— an integer value equal to the number of the reach. IRCH4B must be greater than zero and less than or equal to NREACHES. IRCH4B data can be specified in any order but each entry must be assigned a unique number and be in the range of 1 to NREACHES.

NCONN—an integer value that defines the number of reaches reach IRCH4B is connected to.

ICONN— an integer list equal to the reach numbers connected to reach IRCH4B. ICONN must be in the range of 1 to NREACHES and includes NCONN ICONN entries.

ITEM 4c VARIABLES – Tabular data dimensions:

NTABS— an integer value equal to the number of tabular data items specified in Item 4d.

ITEM 4d VARIABLES – Tabular data:

CTABTYPE— a string value that specifies the data type the selected tabular data item represents.

> CTABTYPE = RAIN, the tabular data item represents rainfall data.
>
> CTABTYPE = EVAP, the tabular data item represents evaporation data.
>
> CTABTYPE = LATFLOW, the tabular data item represents lateral flow data.
>
> CTABTYPE = STAGE, the tabular data item represents stage data for constant stage reaches.
>
> CTABTYPE = STRUCTURE, the tabular data item represents structure data. Structure data can be either structure operation criteria, specified discharge, or gate opening data for operable structures and is defined in Items 13a and 13b.
>
> CTABTYPE = TIME, the tabular data item represents SWR1 time-step length data. CTABTYPE = TIME should only be specified once in Item 4d and overrides SWR1 time step data (RTINI, RTMIN, RTMAX, RTMULT, and NTMULT) specified in Item 2.

ITABUNIT—an non-zero integer value that defines the unit number of the tabular data file. The file unit number ITABUNIT must be specified in the NAM file using the keyword DATA if ITABUNIT is greater than 0 or the keyword DATA(BINARY) if ITABUNIT is less than 0.

CINTP— a string value that specifies the interpolation applied to the selected tabular data item.

CINTP = NONE, the tabular data value closest to but greater than or equal to the end of the current SWR1 time step is used to define the data value for the current SWR1 time step.

CINTP = AVERAGE, the temporally weighted average of all tabular data values from the end of the previous SWR1 time step to the end of the current SWR1 time step is used to define the data value for the current SWR1 time step.

CINTP = INTERPOLATE, the two tabular data values closest to the end of the current SWR1 time step are used to interpolate the data value for the the current SWR1 time step.

CTABRCH— a string variable that is used to define the number of reaches that the tabular data will be applied to. The tabular data can be applied to all reaches in the SWR1 dataset by specifying CTABRCH = ALL. If CTABRCH is not equal to ALL, an integer value must be specified. This variable is only specified for CTABTYPE values other than TIME.

ITABRCH— an integer list equal to the reach numbers that use the tabular data. ITABRCH must be in the range of 1 to NREACHES and includes INT(CTABRCH) ITABRCH entries. This variable is only specified if CTABRCH values other than ALL are specified.

ITEM 5 VARIABLES – Input control for reach data for current stress period:

ITMP— an integer value for reusing SWR1 reach data that can change each stress period. ITMP must be greater than zero for the first stress period of a simulation. For subsequent stress periods, if ITMP is less than or equal to zero, data from the previous stress period will be used.

IRDBND—an integer value that defines the number of reaches where boundary data will be modified during the current MODFLOW stress period. IRDBND must be equal to NREACHES for the first stress period of a simulation. For subsequent stress periods, if IRDBND > 0 boundary data are specified in Item 6; however if IRDBND ≤ 0, boundary values from the previous stress period will be used. For all stress periods, IRDBND must not exceed NREACHES.

IRDRAI—an integer value that defines the number of reaches where rainfall data will be modified during the current MODFLOW stress period. If IRDRAI > 0, reach rainfall data are specified in Item 7a. If IRDRAI < 0, rainfall data are specified for all reaches in Item 7b using two-dimensional arrays and read using U2DREL. For the first stress period, rainfall rates of 0.0 [L/T] are used for all reaches if IRDRAI = 0 or for reaches not defined in Item 7a if IRDRAI > 0. For subsequent stress periods, if IRDRAI = 0 reach rainfall data from the previous stress period will be used for all reaches not using external time series files. For all stress periods, IRDRAI must not exceed NREACHES.

IRDEVP—an integer value that defines the number of reaches where evaporation data will be modified during the current MODFLOW stress period. If IRDEVP > 0, reach evaporation data are specified in Item 8a. If IRDEVP < 0, evaporation data are specified for all reaches in Item 8b using two-dimensional arrays and read using U2DREL. For the first stress period, evaporation rates of 0.0 [L/T] are used for all reaches if IRDEVP = 0 or for reaches not defined in Item 8a if IRDEVP > 0. For subsequent stress periods, if IRDEVP = 0 reach evaporation data from the previous stress period will be used for all reaches not using external time series files. For all stress periods, IRDEVP must not exceed NREACHES.

IRDLIN—an integer value that defines the number of reaches where lateral flow data will be modified during the current MODFLOW stress period. If IRDLIN > 0, reach lateral flow data are specified in Item 9a. If IRDLIN < 0, lateral flow data are specified for all reaches in Item 9b using two-dimensional arrays and read using U2DREL. For the first stress period, lateral flow rates of 0.0 [L³/T] are used for all reaches if IRDLIN = 0 or for reaches not defined in Item 9a if IRDLIN > 0. For subsequent stress periods, if IRDLIN = 0 reach lateral flow data from the previous stress period will be used for all reaches not using external time series files. For all stress periods, IRDLIN must not exceed NREACHES.

IRDGEO—an integer value that defines the number of reaches where geometry data will be modified during the current MODFLOW stress period. If IRDGEO > 0, reach geometry data are specified in Items 10 and 11. IRDGEO must be equal to NREACHES for the first stress period of a simulation. For subsequent stress periods, if IRDGEO ≤ 0, geometry data from the previous stress period will be used for all reaches. For all stress periods, IRDGEO must not exceed NREACHES.

IRDSTR—an integer value that defines the number of reaches where structure data will be modified during the current MODFLOW stress period. If IRDSTR > 0, structure data are specified in Items 12 and 13, however if IRDSTR ≤ 0, structure data from the previous stress period will be used for all reaches. IRDSTR can be zero on the first stress period, indicating that no control structures are present in the model. For all stress periods, IRDSTR must not exceed NREACHES.

IRDSTG—an integer value that defines the number of reaches where stage data will be modified during the current MODFLOW stress period. If IRDSTG > 0, stage data are specified for all reaches in Item 14a. If IRDSTG < 0, stage data are specified for all reaches in Item 14b using a two-dimensional array and read using U2DREL. IRDSTG must be equal to NREACHES or less than zero for the first stress period of a simulation to define an initial stage above sum of the reach bottom (GBELEV) and the minimum reach depth (DMINDPTH). For subsequent stress periods, IRDSTG values not equal to zero only need to be specified for subsequent stress periods if constant-stage reaches with stages different from the previous stress period are defined for the stress period (ISWRBND for reach < 0) or to redefine reach offsets for reservoir-routing reaches. If constant-stage reaches are defined, IRDSTG > 0, and IRDSTG < NREACHES, stage data from the previous stress period will be used for reaches not specified in Item 14a and not using external time series files. For all stress periods, IRDSTG must not exceed NREACHES.

IPTFLG—an integer value for printing input and processed data specified for this stress period. If IPTFLG > 0, input and processed data for this stress period will be printed. If IPTFLG = 0, input and processed data for this stress period will not be printed.

IRDAUX—an optional integer value that defines the number of reaches that auxiliary data will be modified during the current MODFLOW stress period. If IRDAUX > 0 and auxiliary variables are specified in item 1a, auxiliary data are specified in Item 15. IRDAUX only needs to be specified for SWR1 datasets that have auxiliary variables defined in item 1a; IRDAUX is ignored for SWR1 datasets that do not include auxiliary variables. If IRDAUX < 1 or is not specified, auxiliary data are specified in Item 15 and data from the previous stress period will be used for all reaches. IRDAUX must be equal to NREACHES for the first stress period of a simulation for SWR1 datasets with auxiliary variables defined in item 1a. For all stress periods, IRDAUX must not exceed NREACHES.

ITEM 6 VARIABLES – Reach boundary data for current stress period:

IBNDRCH— an integer value equal to the number of the reach. IBNDRCH must be greater than zero and less than or equal to NREACHES. IBNDRCH data can be specified in any order, but each entry must be assigned a unique number.

ISWRBND— an integer value defining the boundary value for reach IBNDRCH. This variable is only specified if IRDBND > 0.

ISWRBND < 0, Constant-stage reach.

ISWRBND = 0, Inactive reach.

ISWRBND > 0, Active reach.

ITEM 7a VARIABLES – Reach rainfall data in list format for current stress period:

IRAIRCH— an integer value equal to the number of the reach. IRCH must be greater than zero and less than or equal to NREACHES. IRCH data can be specified in any order, but each entry must be assigned a unique number.

RAIN—a real value that is the volumetric rate per unit area of water added by precipitation directly onto the reach (in units of length per time).

ITEM 7b VARIABLES – Two-dimensional rainfall data for current stress period:

RAIN2D— an array of positive real values used to define the volumetric rate per unit area of water added by precipitation directly onto the reach (in units of length per time). This array is only specified if IRDEVP < 0.

ITEM 8a VARIABLES – Reach evaporation data in list format for current stress period:

IEVPRCH— an integer value equal to the number of the reach. IEVPRCH must be greater than zero and less than or equal to NREACHES. IEVPRCH data can be specified in any order, but each entry must be assigned a unique number.

EVAP—a real value that is the volumetric rate per unit area of water removed by evaporation directly from the reach (in units of length per time). EVAP is defined as a positive value.

ITEM 8b VARIABLES – Two-dimensional evaporation data for current stress period:

EVAP2D— an array of positive real values used to define the volumetric rate per unit area of water removed by evaporation directly from the reach (in units of length per time). This array is only specified if IRDEVP < 0.

ITEM 9a VARIABLES – Reach lateral flow data in list format for current stress period:

ILINRCH— an integer value equal to the number of the reach. ILINRCH must be greater than zero and less than or equal to NREACHES. ILINRCH data can be specified in any order, but each entry must be assigned a unique number.

QLATFLOW—a real value that is the lateral flow (in units of cubic length per time) entering or leaving the reach. Point inflows and distributed lateral flows must be combined in QLATFLOW. The value can be any number.

ITEM 9b VARIABLES – Two-dimensional lateral flow data for current stress period:

QLATFLOW2D— an array of real values used to define the volumetric rate of water added or subtracted from all reaches in the model grid cell (in units of cubic length per time). Lateral flow data are applied to individual reaches by scaling model grid cell value by the ratio of the length of the reach to the total length of reaches in the model grid cell. This array is only specified if IRDLIN < 0.

ITEM 10 VARIABLES – Reach geometry assignment data in list format for current stress period:

IGMODRCH— an integer value equal to the number of the reach. IGMODRCH must be greater than zero and less than or equal to NREACHES. IGMODRCH data can be specified in any order but each entry must be assigned a unique number.

IGEONUMR— an integer value for defining the geometry data defined in Item 11a associated with the reach. IGEONUMR can change each stress period but must be greater than zero and less than IRDGEO for the first stress period. If IGEONUMR < 1, geometry data for the reach from the previous stress period is reused.

GZSHIFT—a real value that is the vertical offset (in units of length) to apply to geometry data (IGEONUM) associated with the reach and defined in Item 11s. The value can be any number. If IGEONUMR < 1, geometry data for the reach from the previous stress period is reused and GZSHIFT is not used for reach IFMODRCH and GZSHIFT can be any value.

ITEM 11a VARIABLES – Initial or modified geometry data for current stress period:

IGEONUM— an integer value that defines the number of the geometry entry. IGEONUM must be greater than zero and less than or equal to IRDGEO. IGEONUM data can be specified in any order, but each entry must be assigned a unique number. Reaches having IGEONUMR equal to IGEONUM are assigned geometry data for IGEONUM.

IGEOTYPE—an integer value that defines the geometry type.

IGEOTYPE = 1, Rectangular cross section.

IGEOTYPE = 2, Trapezoidal cross section.

IGEOTYPE = 3, Irregular cross section.

IGEOTYPE = 4, Specified stage, volume, wetted perimeter, surface area, cross-sectional area relationship.

IGEOTYPE = 5, Surface-water feature covering the entire finite-difference cell.

IGCNDOP—an integer value that defines the approach used to calculate conductance for the geometry entry.

IGCNDOP = 0, Fixed conductance is specified for the geometry entry.

IGCNDOP = 1, Conductance is calculated using specified leakance coefficient, reach length, and simulated wetted perimeter.

IGCNDOP = 2, Conductance is calculated using the horizontal hydraulic conductivity, reach length, and simulated wetted perimeter.

IGCNDOP = 3, Conductance is calculated using an assumed serial connection (harmonic mean) of the specified leakance coefficient and horizontal hydraulic conductivity.

GMANNING—a real number that is Manning's roughness coefficient for the geometry entry (in units of seconds per m$^{1/3}$ and equivalent to published dimensionless Manning's roughness coefficients). This value is only used for reaches associated with this IGEONUM and using the diffusive-wave approximation (IROUTETYPE = 3) to solve for surface-water flow between reach groups or reaches connected using an uncontrolled discharge connection structure (|ISTRTYPE| = 2).

NGEOPTS—an integer value that defines the number of points used to define the geometry. This variable is only specified if IGEOTYPE = 3 or 4.

GWIDTH—a real number that defines the bottom width of the rectangular or trapezoidal cross-section (in units of length). This variable is only specified if IGEOTYPE = 1 or 2.

GBELEV—a real number that defines the bottom elevation of the rectangular or trapezoidal cross-section (in units of length). This variable is only specified if IGEOTYPE = 1 or 2.

GSSLOPE—a real number that defines the side slope of the trapezoidal cross-section (in units of length per length). This variable is only specified if IGEOTYPE = 2.

GCND—a real number that defines the conductance (in units of length squared per time) of the geometry entry. If a reach associated with this geometry type spans more than one layer (KRCH < 0), GCND is internally distributed based on the length of the wetted perimeter in the layer relative to the total wetted perimeter of the reach. This variable is only specified if IGCNDOP = 0.

GLK—a real number that defines the leakance coefficient (in units of time^{-1}) of the geometry entry. This variable is only specified if IGCNDOP = 1 or 3.

GCNDLN—a real number that defines the average horizontal distance from the reach to the center of the finite-difference grid (in units of length). This variable is only specified if IGCNDOP = 2 or 3.

GETEXTD—a real number that defines the extinction depth (in units of length) for evapotranspiration from groundwater for cases where simulated evaporation for the reach is less than EVAP. This variable is only specified if IGEOTYPE = 5.

ITEM 11b VARIABLES—Irregular cross-section data for geometry type IGEONUM for current stress period:

XB(i)—a real value that is the distance relative to the left bank (in units of length) of the geometry type (when looking downstream). By definition, the first of NGEOPTS values represents the left edge of the cross-section; values for remaining points should be equal to or less than the previous distance.

ELEVB(i)—a real value that is the elevation (in units of length) of the cross-section point at XB(i).

ITEM 11c VARIABLES—Stage-volume-wetted perimeter-surface area-cross-sectional area data for geometry type IGEONUM for current stress period:

ELEV(i)—a real value that is the elevation (in units of length) of the geometry type. By definition, the first of NGEOPTS values represents the lowest elevation; values for remaining points should be greater than the previous elevation.

VOL(i)—a real value that is the volume (in units of length cubed) of the geometry type at ELEV(i). By definition, the first of NGEOPTS values represents the smallest volume (and will likely be zero); values for remaining points should be greater than or equal to the previous volume.

WETPER(i)—a real value that is the wetted perimeter (in units of length) of the geometry type at ELEV(i). By definition, the first of NGEOPTS values represents the smallest wetted perimeter (and will likely be zero); values for remaining points should be greater than or equal to the previous wetted perimeter.

SAREA(i)—a real value that is the surface area (in units of length squared) of the geometry type at ELEV(i). By definition, the first of NGEOPTS values represents the smallest surface area (and will likely be zero); values for remaining points should be greater than or equal to the previous surface area.

XAREA(i)—a real value that is the cross-sectional area (in units of length squared) of the geometry type at ELEV(i). By definition, the first of NGEOPTS values represents the smallest cross-sectional area (and will likely be zero); values for remaining points should be greater than or equal to the previous cross-sectional area. The cross-sectional area is used for conveyance calculations for reaches using the diffusive-wave approximation (IROUTETYPE = 3) to solve for surface-water flow.

ITEM 12 VARIABLES—Reach structure assignment for current stress period:

ISMODRCH—an integer value equal to the number of the reach. ISMODRCH must be greater than zero and less than or equal to NREACHES. ISMODRCH data can be specified in any order, but each entry must be assigned a unique number.

NSTRUCT—an integer value for defining the number of structures associated with the reach. NSTRUCT can change each stress period. If NSTRUCT < 0, structure data from the previous stress period are reused for reach ISMODRCH. If NSTRUCT < 0 on the first stress period, the reach is assumed to have no structures.

ITEM 13a VARIABLES—Initial or modified structure data for current stress period:

ISTRRCH—an integer value equal to the number of the reach. ISTRRCH must be greater than zero and less than or equal to NREACHES. ISTRRCH data can be specified in any order, but each reach identified in Item 12 with a NSTRUCT value greater than zero must be included in Item 13 or structure flow will not be simulated for reach ISTRRCH.

ISTRNUM—an integer value that defines the number of the structure located in reach ISTRRCH. ISTRNUM must be greater than zero and less than or equal to NSTRUCT defined for reach ISMODRCH in Item 12. ISTRNUM data can be specified in any order but a total of NSTRUCT structures should defined for reach ISTRRCH.

ISTRCONN—an integer value equal to the reach number structure ISTRNUM is connected to. ISTRCONN must be less than or equal to NREACHES and consistent with the reach connectivity defined in ITEM 4b (ICONN(1...NCONN)). Values of ISTRCONN equal to zero indicate that the tailwater end of the structure is unconnected.

ISTRTYPE—an integer value that defines the structure type.

ISTRTYPE = -2, Uncontrolled discharge connection structure with a user-specified structure invert elevation. This option is used to define a zero-depth gradient boundary condition for unconnected reaches (ISTRCONN = 0).

ISTRTYPE = 0, No structure.

ISTRTYPE = 1, Specified elevation excess volume structure.

ISTRTYPE = 2, Uncontrolled discharge connection structure. If the reach is unconnected (ISTRCONN = 0), this option is used to define a critical-depth boundary condition.

ISTRTYPE = 3, Pump.

ISTRTYPE = 4, Specified stage-discharge relation.

ISTRTYPE = 5, Culvert.

ISTRTYPE = 6, Fixed crest weir.

ISTRTYPE = 7, Fixed gated spillway (underflow gate).

ISTRTYPE = 8, Movable crest weir (overflow gate).

ISTRTYPE = 9, Gated spillway (underflow gate).

NSTRPTS—an integer value that defines the number of stage-discharge entries in Item 13c. NSTRPTS must be greater than zero but is only specified if ISTRTYPE = 4.

STRWCD—a real number that is a weir discharge coefficient (dimensionless) for the structure entry. This variable is only specified if ISTRTYPE = 5, 6, 7, 8, or 9.

STRWCD2—a real number that is an orifice discharge coefficient (dimensionless) for the structure entry. This variable is only specified if ISTRTYPE = 5, 7, or 9.

STRCD3—a real number that is the dimensionless submergence exponent for the structure. A value of 0.5 is sufficient if specific data are not available for a structure. This variable is only specified if ISTRTYPE = 6, 7, 8, or 9.

STRINV—a real number that is the structure invert elevation (units of length) for the structure entry. This variable is only specified if ISTRTYPE = -2, 5, 6, 7, 8, or 9.

STRINV2—a real number that is the downstream invert elevation (units of length) for the structure entry. This variable is only specified if ISTRTYPE = 5.

STRWID—a real number that is the structure width (units of length) perpendicular to flow for the structure. This variable is only specified if ISTRTYPE = 5, 6, 7, 8, or 9. If ISTRTYPE = 5 and STRWID > 0, the culvert is circular and STRWID is the diameter of the culvert. If ISTRTYPE = 5 and STRWID < 0, the culvert is rectangular and |STRWID| is the culvert span.

STRWID2—a real number that is the culvert rise (units of length). This variable is only specified if ISTRTYPE = 5 and STRWID < 0.

STRLEN—a real number that is the culvert length (units of length). This variable is only specified if ISTRTYPE = 5.

STRMAN—a real number that is the culvert Manning's roughness coefficient. This variable is only specified if ISTRTYPE = 5.

STRVAL—a real number that is the inital flow rate (units of length^3 per time) or gate opening (units of length) for the structure. If ISTRTYPE = 6 or 7, STRVAL is the gate opening for the stress period. This variable is only specified if ISTRTYPE = 3, 6, 7, 8, or 9.

ISTRDIR—an integer value that defines directional limits on flow for the structure. This variable is only specified if ISTRTYPE = 5, 6, 7, 8, or 9.

ISTRDIR < 0, Restriction of flow from ISTRCONN to REACHi.

ISTRDIR = 0, Bi-directional flow is allowed.

ISTRDIR > 0, Restriction of flow from REACHi to ISTRCONN.

ITEM 13b VARIABLES – Initial or modified structure operation data for current stress period:

This item is only specified for structures with ISTRTYPE = 1, 3, 8, or 9.

CSTROTYP—a string variable that is used to set the SWR1 data type used to control structure operations. This variable is only specified if ISTRTYPE = 1, 3, 8, or 9.
CSTROVAL = STAGE, simulated SWR1 stages at a user-defined reach will be used to operate the structure.

CSTROVAL = FLOW, simulated SWR1 flow at a user-defined reach connection will be used to operate the structure. A CSTROVAL = FLOW is not supported for ISTRTYPE = 1.

ISTRORCH—an integer value that defines the reach that will be used to operate the structure. This variable is only specified if ISTRTYPE = 3, 8, or 9.

ISTROQCON—an integer value that defines the reach connection for reach ISTRORCH that will be used to operate the structure. This variable is only specified if ISTRTYPE = 3, 8, or 9 and CSTROVAL = FLOW.

CSTROLO—a string variable that is a logical operation used to determine if a structure should be operated. This variable is only specified if ISTRTYPE = 3, 8, or 9.

CSTROLO = GE, the structure will be operated when the STAGE in reach ISTRORCH or FLOW at the connection between reach ISTRORCH and ISTROQCON is greater than or equal to the specified structure criteria STRCRIT.

CSTROLO = LT, the structure will be operated when the STAGE in reach ISTRORCH or FLOW at the connection between reach ISTRORCH and ISTROQCON is less than the specified structure criteria STRCRIT.

CSTRCRIT—a string variable that is used to set the flag (ISTRTSTYPE) determining if tabular data from an external file will be used to define the control-elevation or control-flow criterion (STRCRIT – units of length or length^3 per time) for the structure. Tabular data from an external file will be used to define STRCRIT if CSTRCRIT begins with the characters TABDATA (for example TABDATA001). CSTRCRIT will be parsed to determine the tabular data item to use to define STRCRIT data for the structure. The tabular data defining STRCRIT must be defined in the tabular data section (Item 4d) and specified in the NAM file using the keyword DATA. Tabular data cannot be used to define the control-elevation control-flow criterion for a structure if tabular data are used to define STRVAL (see the description for CSTRVAL). If CSTRCRIT does not begin with TABDATA, CSTRCRIT will be converted to a real number representing STRCRIT. STRCRIT is the control-elevation control-flow criterion that is evaluated in REACH ISTRORCH to determine if a structure should be operated. This variable is only specified if ISTRTYPE = 1, 3, 8, or 9.

STRCRITC—a real number that is the control offset criterion (units of length or length^3 per time) that must be evaluated in REACH ISTRORCH prior to closing the structure. The closing elevation is calculated by adding STRCRITC and STRCRIT. This variable is only specified if ISTRTYPE = 3, 8, or 9.

STRRT—a real number that is pump startup rate (units of length^3 per time^2) or the opening rate (units of length per time or length^3 per time^2) of the operable structure. This variable is only specified if ISTRTYPE = 3, 8, or 9.

STRMAX—a real number that is the maximum discharge rate (units of length^3 per time) or the maximum opening (units of length) of the structure. This variable is only specified if ISTRTYPE = 1, 3, 8, or 9.

CSTRVAL— a optional string variable that is used to set the flag (ISTRTSTYPE) determining if tabular data from an external file will be used to define the discharge (units of length^3 per time) or gate opening (units of length) for the structure. Tabular data from an external file will be used to define STRVAL if CSTRVAL begins with the characters TABDATA (for example TABDATA001). CSTRVAL will be parsed to determine the tabular data item to use to define STRVAL data for the structure. The tabular data defining STRVAL must be defined in the tabular data section (Item 4d) and specified in the NAM file using the keyword DATA. Tabular data cannot be used to define the gate opening or discharge for a structure if tabular data are used to define STRCRIT (see the description for CSTRCRIT). If CSTRVAL does not begin with TABDATA, the variable is ignored and STRVAL will be calculated using the defined structure operation data. This optional variable is only specified if ISTRTYPE = 1, 3, 8, or 9.

ITEM 13c VARIABLES—Stage-discharge data for the structure for current stress period:

STRELEV(i)—a real value that is the elevation (in units of length) of the structure. By definition, the first of NSTRPTS values represents the lowest elevation; values for remaining points should be greater than the previous elevation. This variable is only specified if ISTRTYPE = 4.

STRQ(i)—a real value that is the flow rate (units of length cubed per time) for the structure entry STRELEV(i). This variable is only specified if ISTRTYPE = 4.

ITEM 14a VARIABLES – Reach stage data for current stress period:

IRCHSTG— an integer value equal to the number of the reach. IRCHSTG must be greater than zero and less than or equal to NREACHES. IRCHSTG data can be specified in any order, but each entry must be assigned a unique number.

STAGE—a real value that is the stage for the reach. On the first stress period, STAGE must be defined for all reaches (NREACHES) with a stage > DMINSTAGE and represents initial stages for the surface-water system. For subsequent stress periods, STAGE only needs to be specified for constant-stage reaches (ISWRBND < 0) or to change the reach offset for reaches using the reservoir-routing approximation (IROUTETYPE = 1 or 2) to simulate surface-water flow; STAGE values are not used to update stages for inactive and active reaches (ISWRBND = 1).

ITEM 14b VARIABLES—Two-dimensional reach stage data for current stress period:

> STAGE2D—an array of positive real values used to define the stage for each reach in model grid cell (in units of length). This array is only specified if IRDSTG < 0. On the first stress period, STAGE2D represents initial stages for the surface-water system. For subsequent stress periods, STAGE2D is used to change the stage in constant-stage reaches (ISWRBND < 0) and/or reach offsets for reaches using the tilted-pool reservoir-routing approximation (IROUTETYPE = 2) to simulate surface-water flow.

ITEM 15 VARIABLES – Reach auxiliary data for current stress period:

> IRCHAUX— an integer value equal to the number of the reach. IRCHAUX must be greater than zero and less than or equal to NREACHES. IRCHAUX data can be specified in any order, but each entry must be assigned a unique number.

> [xyz]—real values representing auxiliary variables for the reach. On the first stress period, [xyz] must be defined for all auxiliary variables defined in item 1a for all reaches. The [xyz] values must be defined in the order used to define the variables in item 1a.

Input Instructions for SWR1 List Utility Subroutine SSWRLSTRD

Subroutine SSWRLSTRD reads lists that are any number of repetitions of an input item that contains multiple variables. SWR1 input items that use this subroutine are items 4, 6, 7a, 8a, 9a, 10, 11b, 11c, 12, 13c, and 14a.

1. [CTAG] [IUNIT] [CFNAME]

Item 1 is optional.

2. LIST

Explanation of variables read by the SWR1 List Utility Subroutine:

> CTAG—a text keyword defining the location of the list for the specific input item. Valid values are INTERNAL, EXTERNAL, or OPEN/CLOSE. If CTAG is not specified, the list is read from the SWR1 input file and is equivalent to specifying CTAG to be INTERNAL.

> IUNIT—an integer value that defines the unit number for the file containing the list for the specific input item. The name of the file associated with this file unit must be contained in the Name File, and its file type must be 'DATA' in the Name File. IUNIT is only specified if CTAG is EXTERNAL.

> CFNAME—a text value that defines the name of the file containing the list for the specific input item. This file will be opened on unit 99 just before reading the list and closed immediately after the list is read. This file should not be included in the Name File. CFNAME is only specified if CTAG is OPEN/CLOSE.

> LIST—is the specified number of lines of data in which each line contains a specified number of variables. The number of fields varies according to which input item is calling this routine. For example, two fields (IRCHSTG and STAGE) are read in input item 14a.

Input Instructions for SWR1 Time Series Utility Subroutine SSWR_RDTABDATA

Subroutine SSWR_RDTABDATA reads time series data from external ASCII (formatted) and binary (unformatted) files and can be used to define rainfall (RAIN), evaporation (EVAP), lateral flow (QLATFLOW), constant-stage reach stages (STAGE), structure data (STRCRIT or STRVAL) for operable surface water control structures, and SWR1 time-step lengths. The SWR1 input item that uses this subroutine is item 4d.

0. [#Text]

Item 0 is optional—# must be in column 1. Item 0 can be repeated multiple times.

1. TIMEOFFSET TIMEOFFSETVAL TIMESCALE TIMESCALEVAL OFFSET OFFSETVAL SCALE SCALEVAL

Item 1 is optional.

2. SIMTIME(1) VALUE(1)

```
SIMTIME(2)  VALUE(2)
```

.

.

.

```
SIMTIME(N)  VALUE(N)
```

Explanation of variables read by the SWR1 Time Series Utility Subroutine:

TIMEOFFSETVAL—an optional real value that defines the offset to apply to the simulation time (SIMTIME) specified in item 2. If TIMEOFFSETVAL is defined, it must be preceeded by the keyword TIMEOFFSET. If TIMEOFFSETVAL is not defined, a default value of 0.0 is used. The SIMTIME used by the SWR1 Process is the product of TIMESCALEVAL (defined below) and sum of SIMTIME and TIMEOFFSETVAL. TIMEOFFSETVAL can be used to modify the time for a dataset (for example, convert SIMTIME data from one relative starting point to another).

TIMESCALEVAL—an optional real value that defines the scaling factor to apply to the adjusted SIMTIME (the sum of SIMTIME and TIMEOFFSETVAL) derived from the values specified in item 2. If TIMESCALEVAL is defined, it must be preceeded by the keyword TIMESCALE. If TIMESCALEVAL is not defined, a default value of 1.0 is used. The SIMTIME used by the SWR1 Process is the product of TIMESCALEVAL and the sum of SIMTIME and TIMEOFFSETVAL (defined above). TIMESCALEVAL can be used to perform a unit conversion on the SIMTIME in a dataset (for example, convert SIMTIME data from hours to days).

OFFSETVAL—an optional real value that defines the offset to apply to the values (VALUE) specified in item 2. If OFFSETVAL is defined, it must be preceeded by the keyword OFFSET. If OFFSETVAL is not defined, a default value of 0.0 is used. The value used by the SWR1 Process is the product of SCALEVAL (defined below) and the sum of VALUE and OFFSETVAL. OFFSETVAL can be used to modify the datum for a dataset (for example, convert stage data from one vertical datum to another).

SCALEVAL—an optional real value that defines the scaling factor to apply to the adjusted values (the sum of VALUE and OFFSETVAL) derived from the values specified in item 2. If SCALEVAL is defined, it must be preceeded by the keyword SCALE. If SCALEVAL is not defined, a default value of 1.0 is used. The value used by the SWR1 Process is the product of SCALEVAL and the sum of VALUE and OFFSETVAL (defined above). SCALEVAL can be used to perform a unit conversion on the values in a dataset (for example, convert stage data from feet to meters).

SIMTIME—a real value that defines the time of the VALUE specified on the same line. Times are relative to the model simulation time (TOTIM). The last line of the must have a SIMTIME value that exceeds the total model simulation time.

VALUE—a real value that defines the value at the select simulation time for the data type specified in Item 4d. The units must be consistent with the data type (CTABTYPE) specified in Item 4d.

For binary (unformatted) external time series files, Items 0 and 1 must not be included and variables in Item 2 must be written as single precision variables.

Appendix 3.   Numerical Solution of Surface-Water Routing Applied in the SWR1 Process

The Q_M term in equation 17 is nonlinear (as a result of equation 9) except for cases where the reservoir-routing approximation is used with either a linear control-structure equation or a linear rating curve. Furthermore, except for rectangular reaches, Q_{PR}, Q_{EV}, Q_{AQ}, Q_{CS}, and ∂V are a function of the stage, h, at the current $(k+1)$ and previous (k) time steps. A typical approach to solving initial-value problems like equation 17 is to (1) linearize the stage-dependent terms, (2) solve the linearized system of equations using direct or iterative methods, and (3) use outer (Picard) iterations to refine the nonlinear contributions (Picard, 1890). An alternative, but equally valid, approach to solving equation 17 is to use a Newton approach that involves use of nonlinear Newton iterative methods within outer iterations to linearize a problem with subsequent use of direct or iterative methods to solve the linearized problem. The advantages of the Newton approach include its ability to solve nonlinear problems, like equation 17, and the superlinear to asymptotic quadratic convergence properties of the approach (Knoll and Keyes, 2004).

To permit the greatest flexibility to solve for surface-water flow for a variety of settings and geometries, Newton approaches are used in the SWR1 Process to solve a system of equations developed using a residual form of equation 17 (and potentially including the finite difference form of Q_M shown in equation 18 and/or the general structure shown in equation 19). The residual form of a system of equations based on equation 17 in vector form is

$$\mathbf{F}(\mathbf{h}) \quad \frac{\Delta \mathbf{V}}{\Delta t} + \sum_{i=1}^{nconn} \mathbf{Q}_M + \mathbf{Q}_{PR} + \mathbf{Q}_{LAT} \quad \mathbf{Q}_{EV} + \mathbf{Q}_{AQ} + \mathbf{Q}_{CS},$$

(3–1)

where

$\quad \mathbf{F}(\mathbf{h}) \qquad$ is the vector-valued function of nonlinear residuals (solution to system of continuity equations), and

$\quad \mathbf{h} \qquad$ is the SWR1 stage vector to be found.

Applying Newton's method (Newton, 1671) for $\mathbf{F}(\mathbf{h}) \quad 0$ and ignoring second order and higher terms

$$\mathbf{F}(\mathbf{h}^{k+1}) \quad \mathbf{F}(\mathbf{h}^k) + \mathbf{F}'(\mathbf{h}^k)(\mathbf{h}^{k+1} \quad \mathbf{h}^k).$$

(3–2)

where

$\quad k \qquad$ is the nonlinear iteration index, and

$\quad \mathbf{F}'(\mathbf{h}) \qquad$ is the derivative of $\mathbf{F}(\mathbf{h})$.

Setting the right-hand side of equation 3–2 to zero yields a strict Newton method; iteration of a linear system of equations yields

$$\mathbf{J}(\mathbf{h}^k)\mathbf{v}^k \quad \mathbf{F}(\mathbf{h}^k), \; \mathbf{h}^{k+1} \quad \mathbf{h}^k + \mathbf{v}^k, \; k \quad 0,1,\ldots$$

(3–3)

where

$\quad \mathbf{J} \qquad$ is the Jacobian matrix, and

$\quad \mathbf{v} \qquad$ is a calculated upgrade vector.

The Jacobian matrix is equal to $\mathbf{F}'$ in equation 3–2 and defines the response of the vector-valued function $\mathbf{F}(\mathbf{h})$ to a change in the state vector. For two coupled nonlinear equations $F_1(h_1, h_2) \quad 0$, $F_2(h_1, h_2) \quad 0$ the Jacobian matrix is

$$\mathbf{J} \quad \begin{bmatrix} \dfrac{\partial F_1}{\partial h_1} & \dfrac{\partial F_1}{\partial h_2} \\ \dfrac{\partial F_2}{\partial h_1} & \dfrac{\partial F_2}{\partial h_2} \end{bmatrix}.$$

(3–4)

The Jacobian can be formulated analytically from the governing equations or approximated using a first-order or second-order finite-difference evaluation of equation 17. A first-order approximation of the Jacobian is

$$\mathbf{J}\left(\mathbf{h}^k\right) \approx \tilde{\mathbf{J}}\left(\mathbf{h}^k\right) \quad \frac{\left[\mathbf{F}\left(\mathbf{h}^k + \sigma\right) \quad \mathbf{F}\left(\mathbf{h}^k\right)\right]}{\sigma},$$

(3–5)

where

$\tilde{\mathbf{J}}$ is the approximate Jacobian, and
σ is an appropriately chosen perturbation.

An approximate Jacobian is computationally expensive to formulate because it requires two evaluations for each simultaneous equation being solved; however, it is the best approach in cases where analytical derivatives are difficult or impossible to formulate. Because equation 17 can be highly nonlinear, an approximate Jacobian is used in the SWR1 Process.

The error in the approximate Jacobian is proportional to σ and is sensitive to scaling, given $\mathbf{h}$ and $\mathbf{v}$. If σ is too large, the derivative is poorly approximated, and if it is too small, the result of the finite difference will be affected by floating-point roundoff error. The best σ to use is one that achieves accuracy by balancing estimates of floating-point error and truncation error. The perturbation calculated in SWR1 is based on Kelley (2003) and is defined as

$$\sigma \quad \max\left(\sqrt{\varepsilon_{mach}}\left|\mathbf{h}^k\right|, \sqrt{\varepsilon_{mach}}\right),$$

(3–6)

where ε_{mach} is the machine epsilon (or unit roundoff).

The advantage of the perturbation defined in equation 3–6 is that it scales the perturbation based on the magnitude of the current solution ($\mathbf{h}$), and ensures that it is no smaller than the square root of the machine epsilon. Kelley (2003) indicates that scaling of the perturbation makes little difference in most cases but can be crucial if $\left|h^k\right|$ is very large.

Another advantage of using an approximate Jacobian is that it can be assembled programmatically using the nonlinear finite-difference equations (for example, equation 18) directly rather than using a linearized form of the equations (for example, a finite-difference form of equation 10) as is done with traditional linear methods. Programmatic evaluation of the approximate Jacobian also facilitates easy incorporation of alternative governing equations and additional source/sink terms.

The linear system $\tilde{\mathbf{J}}\left(\mathbf{h}^k\right)\mathbf{v}^k \quad \mathbf{F}\left(\mathbf{h}^k\right)$ is solved using either lower-upper (LU) decomposition or Krylov methods. LU decomposition results in an exact solution of the linear system whereas Krylov methods result in an approximate solution of the linear system (Krylov, 1931). In general, LU decomposition should result in a better solution during each outer iteration, but will have a higher computational cost. Although Krylov methods result in an approximate solution of the linear system, they are the most practical for large problems (typically greater than 1,000 active reach groups).

Specifically, LU decomposition solves the linear system of equations defined in equation 3–3 by

$$\mathbf{J}\left(\mathbf{h}^k\right)\mathbf{v}^k \quad \left(\mathbf{LU}\right)\mathbf{v}^k \quad \mathbf{L}\left(\mathbf{Uv}^k\right) \quad \mathbf{F}\left(\mathbf{h}^k\right),$$

(3–7)

where

$\mathbf{L}$ is the lower triangular matrix ($a_{n,m} = 0$ for $n < m$), and
$\mathbf{U}$ is the upper triangular matrix ($a_{n,m} = 0$ for $n > m$).

The system of linear equations is solved by first solving for the vector $\mathbf{y}$ such that

$$\mathbf{Ly} \quad \mathbf{F}\left(\mathbf{h}^k\right)$$

(3–8)

and then solving

$$\mathbf{Uv}^k = \mathbf{y}.$$

(3–9)

LU decomposition in SWR1 is performed using Crout's algorithm (Crout, 1941) to solve the set of $nrg^2 + nrg$ linear equations, where nrg is the number of linear equations solved and equal to the number of active reach groups, for the elements of the lower ($\mathbf{L}$) and upper ($\mathbf{U}$) triangular matrices in a defined operation order (Press and others, 1990). LU decomposition is solved using either a full nrg x nrg matrix or a reduced nrg x nrg_r matrix, where nrg_r is 1 plus 2 times the maximum distance between the diagonal and connected reach groups in each linear equation (row). An example of the connectivity of $\tilde{\mathbf{J}}$ for a hypothetical 6 reach group problem is

$$
\tilde{\mathbf{J}} \quad
\begin{bmatrix}
D_{1,1} & a_{1,2} & - & - & - & - \\
a_{2,1} & D_{2,2} & a_{2,3} & - & a_{2,5} & - \\
- & a_{3,2} & D_{3,3} & a_{3,4} & - & - \\
- & - & a_{4,3} & D_{4,4} & a_{4,5} & a_{4,6} \\
- & a_{5,2} & - & a_{5,4} & D_{5,5} & a_{5,6} \\
- & - & - & a_{6,4} & a_{6,5} & D_{6,6}
\end{bmatrix},
\tag{3-10}
$$

where

$D_{n,n}$ is the diagonal of reach group n,

$a_{n,m}$ is the off-diagonal element (column) of reach group n,

-     is an unconnected element (column) of reach group n, and

nrg_r would be equal to $7 = (5\text{-}2) \times 2 + 1$.

The reduced nrg x nrg_r matrix is used to solve the linear equation using LU decomposition in cases where $nrg_r \le nrg/2$. In equation 3–10, the connectivity defines the locations of the off-diagonal elements and as a result an off-diagonal element $a_{n,m}$ (for example $a_{4,3}$) would have a corresponding off-diagonal element at $a_{m,n}$ (for example $a_{3,4}$). Although the connectivity determines the structure of $\tilde{\mathbf{J}}$, the matrix can be nonsymmetric ($a_{n,m} \ne a_{m,n}$), because σ_n may not equal σ_m and may result in conditions where $\partial F_{n,m} \ne \partial F_{m,n}$.

Krylov methods differ from LU decomposition in that only matrix-vector products are required to carry out each inner iteration of the linear solver. Typically, constants are computed by taking inner products (dot products) of residuals or other vectors arising from the iterative method. Krylov methods are projection methods for solving the linear system $\tilde{\mathbf{J}}(\mathbf{h}^k)\mathbf{v}^k \quad \mathbf{F}(\mathbf{h}^k)$ using the Krylov subspace, $\mathbf{K}^{ir}$:

$$
\mathbf{K}^{ir} \quad \text{span}\left(\mathbf{r}_0, \tilde{\mathbf{J}}\mathbf{r}_0, \tilde{\mathbf{J}}^2\mathbf{r}_0, ..., \tilde{\mathbf{J}}^{ir\ 1}\mathbf{r}_0\right),
\tag{3-11}
$$

where

ir is the Krylov iteration number, and

$\mathbf{r}_0$ are the initial residuals calculated using initial stages, $\mathbf{h}_0$

Because the vast majority of fully coupled nonlinear water-resource applications result in Jacobian matrices that are non-symmetric, a biconjugate gradient stabilized (Bi-CGSTAB) and a restarted generalized minimal residual (GMRESm) method have been included in SWR1. Krylov methods are derived either from the long-recurrence Arnoldi orthogonalization procedure (Arnoldi, 1951), which generates orthonormal bases of the Krylov subspace, or the short recurrence Lanczos bi-orthogonalization procedure (Lanczos, 1950), which generates non-orthogonal bases (Knoll and Keyes, 2004). The Bi-CGSTAB and GMRESm methods in SWR1 were developed using Barrett and others (1994) and are briefly described as follows:

• The BiCGSTAB method applies updating operations to a vector based on a system with the original coefficient matrix $\tilde{\mathbf{J}}$ and $\tilde{\mathbf{J}}^T$ to improve the convergence rate. Instead of orthogonalizing each sequence, they are made mutually orthogonal, or "bi-orthogonal."

• The GMRESm method computes a sequence of orthogonal vectors and combines these through a least-squares solve-and-update process. Computational and storage costs are reduced by specifying a fixed number of vectors to be generated and restarting the process with the most recent iterate.

The convergence rate of Krylov methods depends on the spectral properties of $\tilde{\mathbf{J}}$. Preconditioning of $\tilde{\mathbf{J}}$ is often used to transform the linear system into one that is equivalent, in the sense that it has the same solution $\mathbf{h}^{k+1}$, but has more favorable spectral properties and requires less work to obtain a solution (Barrett and others, 1994). For example, if a matrix $\mathbf{M}$ approximates $\tilde{\mathbf{J}}$ in some way, the transformed system

$$\mathbf{M}^{-1}\mathbf{J}\left(\mathbf{h}^{k}\right)\mathbf{v}^{k} = \mathbf{M}^{-1}\mathbf{F}\left(\mathbf{h}^{k}\right) \tag{3-12}$$

has the same solution as the original system $\tilde{\mathbf{J}}\left(\mathbf{h}^{k}\right)\mathbf{v}^{k} = \mathbf{F}\left(\mathbf{h}^{k}\right)$, but the spectral properties of its coefficient matrix $\mathbf{M}^{-1}\tilde{\mathbf{J}}$ may be more favorable. Setup and application of the preconditioner $\mathbf{M}$ adds some extra computational cost; as a result, there is tradeoff between the extra computational cost and the gain in convergence speed. Furthermore, there is no guarantee that use of a preconditioner will improve convergence speed. As a result, four preconditioners have been added to SWR1 to provide options for preconditioning $\tilde{\mathbf{J}}$ and include

- A Jacobi preconditioner where $\mathbf{M}$ is equal to $1/D_{n,n}$ where $D_{n,n}$ are the diagonal elements of $\tilde{\mathbf{J}}$ (Barrett and others, 1994).

- A zero fill in incomplete LU (ILU(0)) factorization preconditioner based on Saad (2003). ILU(0) is the simplest form of ILU preconditioners and has storage requirements identical to $\tilde{\mathbf{J}}$. The accuracy of ILU(0) may be insufficient to yield an adequate rate of convergence to compensate for the increased cost of formulating the preconditioner $\mathbf{M}$. In these cases, more accurate ILU factorizations that compensate for dropped terms (MILU(0)) or allow some additional fill-in (ILUT) for the preconditioner $\mathbf{M}$ may be required.

- A zero fill in modified incomplete LU (MILU(0)) factorization preconditioner based on Saad (2003). MILU(0) is a variant of the ILU(0) preconditioners and has storage requirements identical to $\tilde{\mathbf{J}}$. Elements that are dropped during the incomplete elimination process are compensated for in MILU(0) by summing the dropped entries and subtracting this sum from the diagonal entry. This approach is equivalent to MIC preconditioning applied to symmetric matrices and available in the PCG Package (Hill, 1990).

- A level fill in incomplete LU with dual threshold drop strategy (ILUT) factorization preconditioner based on Saad (1994a) from the SPARSKIT library (Saad, 1994b). ILUT is an incomplete LU factorization that allows the user to determine the level-of-fill and tolerance to be evaluated for dropping small elements. The storage requirements and additional computational cost of the ILUT preconditioner exceeds the storage required for the ILU(0) and MILU(0) preconditioners but can significantly reduce the number of iterations required for convergence.

Unpreconditioned Krylov methods can also be used in SWR1. In cases where preconditioning is not used, $\mathbf{M}$ is identical to the identity matrix $\mathbf{I}$. Practically, the Krylov methods included in SWR1 should be implemented with and without preconditioning to determine if preconditioning improves convergence speeds.

An inexact Newton condition can be used with the Krylov methods in SWR1 to prevent oversolving the Newton equation (USE_INEXACT_NEWTON option). Oversolving the Newton equation may result in little to no progress towards a solution and may involve unnecessary computational expense. As a result, a less accurate solution of equation 3–3 may be computationally less costly and more effective in reducing the ℓ^{2} norm of $\mathbf{F}$ (Eisenstat and Walker, 1996; Pernice and Walker, 1998). The Newton step for the Krylov methods are terminated when

$$\left\|\mathbf{F}\left(\mathbf{h}^{ir}\right)\right\|_{2} \le \eta^{k}\left\|\mathbf{F}\left(\mathbf{h}^{k}\right)\right\|_{2}, \quad ir = 1,2,...,ir_{\max}, \tag{3-13}$$

where

$\left\|\mathbf{F}\left(\mathbf{h}^{ir}\right)\right\|_{2}$ is the ℓ^{2} norm of the current residual,

η^{k} is a forcing function,

$\left\|\mathbf{F}\left(\mathbf{h}^{k}\right)\right\|_{2}$ is the ℓ^{2} norm at the beginning of the inner iteration, and

$ir_{\max}$ is the user-specified maximum number of Krylov iterations (NINNER).

The ℓ^{2} norm of the current residual is

$$\left\|\mathbf{F}\left(\mathbf{h}^{ir}\right)\right\|_{2} = \sqrt{\mathbf{F}\left(\mathbf{h}^{ir}\right)^{T}\mathbf{F}\left(\mathbf{h}^{ir}\right)}. \tag{3-14}$$

The ℓ^2 norm at the beginning of the inner iteration is calculated in the same manner as equation 3–14 except $\mathbf{F}\left(\mathbf{h}^{k}\right)$ is used instead of $\mathbf{F}\left(\mathbf{h}^{ir}\right)$. The forcing function used in SWR1 is

$$\eta^{k} = \gamma \max\left(\left(\frac{\left\|\mathbf{F}\left(\mathbf{h}^{k}\right)\right\|_{2}}{\left\|\mathbf{F}\left(\mathbf{h}^{k-1}\right)\right\|_{2}}\right)^{2}, \left(\eta^{k-1}\right)^{2}\right), \quad k = 1,2,...,k_{max},$$

(3–15)

where

γ is a dampening parameter that is set at 0.9 in SWR1,

$\left\|\mathbf{F}\left(\mathbf{h}^{k}\right)\right\|_{2}$ is the ℓ^2 norm of the residual at the start of the current linear iteration (k),

$\left\|\mathbf{F}\left(\mathbf{h}^{k-1}\right)\right\|_{2}$ is the ℓ^2 norm at the start of the previous linear iteration (k-1),

η^{k-1} is the forcing function from the previous outer iteration, and

k_{max} is the is the user-specified maximum number of nonlinear iterations (NOUTER).

The forcing function used in SWR1 reflects the amount of decrease between the function evaluated at the current nonlinear solution iterate and the function at the previous iterate, and scales the tolerance used to terminate the Krylov solvers based on the relative distance from the true solution of $\mathbf{u}$ (Eisenstat and Walker, 1996; Jones and Woodward, 2001). Close to the solution of $\mathbf{u}$, the forcing function results in a low tolerance, and enforces precise solutions of the linear system of equations. Further away from the solution of $\mathbf{h}$, the tolerance is high and prevents oversolving of the linear system of equations when a highly precise approximation of the Jacobian system will not provide more value than a coarse approximation of the nonlinear system solution.

The storage requirements and computational cost of forming solution iterates using Krylov methods can be further reduced using sparse-matrix storage formats. A modified compressed row storage (MCRS) format that only stores the non-zero elements of $\tilde{\mathbf{J}}$ is used in SWR1. In the MCRS format, the non-zero elements of $\tilde{\mathbf{J}}$ are stored along with pointers to the column (*JA*) for each non-zero element of $\tilde{\mathbf{J}}$, and pointers to the first storage element for each active reach group (*IA*). To facilitate formulation of the compressed preconditioned matrix, $\mathbf{M}$, that approximates the coefficient matrix, $\tilde{\mathbf{J}}$, pointers to the position of the first upper triangular storage element for each reach group (*IU*) are also stored. The MCRS format is a variant of the standard compressed row storage format in which the diagonal is stored in the first storage element for each active reach group. The MCRS format for the hypothetical six-reach group problem shown in equation 3–10 is

$$\begin{bmatrix} JA \\ IA \\ IU \\ \tilde{\mathbf{J}} \end{bmatrix} = \begin{bmatrix} 1 & 2 & 2 & 1 & 3 & 5 & 3 & 2 & 4 & 4 & 3 & 5 & 6 & 5 & 2 & 4 & 6 & 6 & 4 & 5 \\ 1 & & 3 & & & 7 & & & 10 & & & & 14 & & & & 18 & & & 21 \\ 2 & & 5 & & & 9 & & & 12 & & & & 17 & & & & 21 & & & \\ D_{1,1} & a_{1,2} & D_{2,2} & a_{2,1} & a_{2,3} & a_{2,5} & D_{3,3} & a_{3,2} & a_{3,4} & D_{4,4} & a_{4,3} & a_{4,5} & a_{4,6} & D_{5,5} & a_{5,2} & a_{5,4} & a_{5,6} & D_{6,6} & a_{6,4} & a_{6,5} \end{bmatrix}.$$

(3–16)

The sizes of *JA* and $\tilde{\mathbf{J}}$ are equal to the number of non-zero elements in $\tilde{\mathbf{J}}$, which is a function of the connectivity of active reach groups, the size of IA is equal to the number of active reach groups + 1, and the size of IU is equal to the number of active reach groups.

The iterative algorithm used to solve equation 3–3 is achieved using an iterative process that uses LU decomposition or Krylov methods to solve the linearized system of equations is

Let $\mathbf{h}^0$ be given.

For k 1 to k_{max} until "Outer CONVERGENCE"

   Formulate $\tilde{\mathbf{J}}\left(\mathbf{h}^k\right)$

   Formulate $\mathbf{M}\left(\mathbf{h}^k\right)$ from $\tilde{\mathbf{J}}\left(\mathbf{h}^k\right)$

   $\mathbf{M}\left(\mathbf{h}^k\right)$ $\tilde{\mathbf{J}}\left(\mathbf{h}^k\right)$ for LU decomposition or unpreconditioned

   Use LU decomposition or Krylov methods:

      For i   1 to ir_{max} until "Inner CONVERGENCE"

         Solve linear system $\tilde{\mathbf{J}}\left(\mathbf{h}^k\right)\mathbf{v}^k$ $\mathbf{F}\left(\mathbf{h}^k\right)$

   Apply dampening:

      $\mathbf{v}^k$ $\mu\mathbf{v}^k$, $0 < \mu \leq 1$

   Update $\mathbf{h}$:

      $\mathbf{h}^{k+1}$ $\mathbf{h}^k + \lambda\mathbf{v}^k$, $0 < \theta \leq 1$

      λ minimizes $\left\|\mathbf{F}^{k+1}\right\|_2$ using Brent's method when $\left\|\mathbf{F}^{k+1}\right\|_2 > \left\|\mathbf{F}^k\right\|_2$ (3–17)

where

   μ is a user-specified dampening/acceleration factor, and

   λ is a line-search scalar calculated using Brent's method (Press and others, 1990) to minimize the ℓ^2 norm of the residual ($\mathbf{F}$) for the current outer iterations ($iter_{outer}$).

Newton's method defined in equation 3–3 can diverge if the initial value is not sufficiently close to the root. Dampening and/or globalization strategies are often employed to ensure that some progress is made towards a solution during each outer iteration ($iter_{outer}$) from almost any starting point. Dampening and a line-search method have been included in SWR1 to facilitate solution of equation 3–1 using the algorithm defined in equation 3–17.

When dampening is applied, the Newton update vector, $\mathbf{v}$, is multiplied by the user-specified scalar μ after convergence is achieved using the selected linear solution method or the maximum number of inner iterations is reached. There is no guarantee that the resultant $\mathbf{v}$ will lead to a reduction in the ℓ^2 norm, but dampening can be an acceptable strategy in cases where initial stage estimates are inaccurate and/or internal source/sink terms ($\mathbf{Q}_{LAT}$, $\mathbf{Q}_P$, $\mathbf{Q}_{ET}$, $\mathbf{Q}_{AQ}$) are unbalanced. These conditions can commonly occur during the first SWR1 time step or during steady-state MODFLOW stress periods.

In the line-search method, the Newton update vector, $\mathbf{v}$, is assumed to be a good direction to move but is multiplied by a scalar, λ, to minimize the ℓ^2 norm of the residual, $\mathbf{F}$; the line-search method can be thought of as dynamic dampening controlled by the ℓ^2 norm of the residual. In SWR1, $0 > \lambda \geq 1$ are evaluated to determine the λ value that results in the smallest ℓ^2 norm. The line-search method in SWR1 should only be used in cases where achieving convergence is difficult because the line-search method increases the computational cost of each outer iteration. Because the line-search method only requires additional evaluations of equation 3–1 and application of the scalar λ applied to $\mathbf{h}$, the computational cost is relatively small. As a result, it is expected the benefits of applying the line-search method will outweigh the computational cost for many applications of SWR1.

SWR1 node reordering using the reverse Cuthill-McKee reordering algorithm available in the Fortran 90 reverse Cuthill-McKee ordering library developed by Burkardt (2011) has been included as an option in the SWR1 Process (USE_RCMREORDERING option). An option for using reverse Cuthill-McKee reordering only if it reduces the bandwidth has also been included (USE_RCMREORDERING_IF_IMPROVEMENT option). Reverse Cuthill-McKee reordering is an effective method for reducing the bandwidth by reordering node numbers using the Jacobian connectivity and maximizing the product of diagonal entries. Jacobian scaling has also been included in the SWR1 Process to improve the condition number of the Jacobian. A symmetric diagonal scaling method (USE_DIAGONAL_SCALING option), equivalent to the matrix scaling method available in the PCG solver of Hill (1990), and scaling using the ℓ^2 norm (USE_L2NORM_SCALING option) have been included. For some problems, the use of reverse Cuthill-McKee reordering and/or Jacobian scaling may improve convergence of the linear solver.

Example Application of Newton's Method in the SWR1 Process

A one-dimensional surface-water problem with a constant inflow and constant downstream stage is used to demonstrate key steps of the application of the Newton approach implemented in the SWR1 Process to solve equation 17. The problem is a steady-state, "surface-water-only" problem (ISWRONLY = 1) consisting of four sequentially connected reaches with the properties summarized in table 3-1 and solved using LU decomposition (ISOLVER = 1). In this problem, reaches and reach groups are equivalent. A flow residual (TOLR) convergence criteria of 1×10^{-9} m³/s was specified and the maximum number of SWR1 nonlinear iterations (NOUTER) was specified to be 50. A steady-state dampening factor (DAMPSS) of 1.0 was used and the line-search method (IBT = 0) was not used. The specified TOLR is much smaller than would be specified for most practical applications but is used here to maximize the number of SWR1 nonlinear iterations for demonstration purposes. Furthermore, for most practical applications with at least one steady-state stress period, a DAMPSS value less than one would be specified. Results from this example application are reported using double-precision numbers to provide enough precision for readers to work through the calculations by hand with minimal round-off error.

Table 3–1.   SWR1 reach parameters for example application of Newton's method in the SWR1 Process.

Reach	Geometry type	Reach length, in meters	Initial stage, in meters	Width, in meters	Bottom elevation, in meters	Routing type	Lateral inflow, in cubic meters per second
1	Rectangular	1,000.0	1.51	20.0	0.30	Diffusive-wave	6.0
2	Rectangular	1,000.0	1.40	20.0	.20	Diffusive-wave	.0
3	Rectangular	1,000.0	1.27	20.0	.10	Diffusive-wave	.0
4	Rectangular	1,000.0	1.00	20.0	.0	Constant stage	.0

The active reach connectivity for this example problem is

$$\begin{bmatrix} D_{1,1} & a_{1,2} & - \\ a_{2,1} & D_{2,2} & a_{2,3} \\ - & a_{3,2} & D_{3,3} \end{bmatrix}.$$

$$(3–M1)$$

The first step in a SWR1 nonlinear iteration is evaluation of the residual (equation 3–1) for each active reach using the initial stage at the beginning of the SWR1 nonlinear iteration. For the initial SWR1 nonlinear iteration, the residual is calculated using the initial heads summarized in table 3–1. The residual for each active reach is evaluated by

1. Determining the volume change using the change in stage from the last to current time-step ($\frac{\Delta V}{\Delta t}$ term in equation 3–1 for the current reach).

2. Determining the sum of the specified lateral flow and current estimates of excess precipitation and groundwater discharge calculated by the UZF1 Package for each reach (Q_{LAT} term in equation 3–1 for the current reach).

3. Determining the surface area and wetted perimeter at the current stage and groundwater head to calculate the Q_{PR}, Q_{EV}, and Q_{AQ} terms in equation 3–1. For the initial SWR1 nonlinear iteration of each MODFLOW time step, the initial stage and groundwater head or stage and groundwater head at the end of the last time step are used.

4. Determining the cross-sectional area for each reach connected to the current reach and equation 10 to calculate the Q_M term in equation 3–1 for the current reach using the same stage and groundwater head used in the previous step.

Q_{CS} is equal to zero for active reaches and equal the sum of all internal sources/sinks (lateral flow, rainfall, evaporation, and aquifer-reach flow) and lateral flows to and from connected reaches for constant stage reaches. Because this problem is a steady-state problem, $\frac{\Delta V}{\Delta t}$ is zero.

The cell-interface components (cross-sectional flow area, nonlinear diffusion coefficient, and stage gradient) used to calculate the discharge, Q_M, for each reach in equation 3–1 and resultant Q_M values using initial stages in table 3–1 for SWR1 nonlinear iteration $k = 0$ are

$$\left(\mathbf{A} \ \middle| \ \mathbf{D}_M \ \middle| \ \left(\frac{\Delta \mathbf{h}}{\Delta \mathbf{x}}\right) \ \middle| \ \mathbf{Q}_M \right) \quad \begin{bmatrix} 24.10000 & 1000.81384 & 0.00011000 & 2.65316041594328 \\ 23.70000 & 911.47656 & 0.00013000 & 2.80828769333504 \\ 21.70000 & 599.73196 & 0.00027000 & 3.51509966363854 \end{bmatrix}. \tag{3–M2}$$

Using function calls for the specified inflow in reach 1 and Q_M terms in equation 3–M2, the resultant residual, $\mathbf{r}$, for SWR1 nonlinear iteration $k = 0$ in each active reach is

$$\begin{bmatrix} 6.00 & 2.65316041594328 \\ 2.65316041594328 & 2.80828769333504 \\ 2.80828769333504 & 3.51509966363854 \end{bmatrix} \quad \begin{bmatrix} 3.34683958405672 \\ 0.155127277391763 \\ 0.706811970303500 \end{bmatrix}. \tag{3–V1}$$

The $\dfrac{\Delta V}{\Delta t}$, Q_{PR}, Q_{ET}, and Q_{AQ} terms are zero for this problem, but would be evaluated with similar function calls to calculate the residual for each reach in cases where these terms are non-zero. The unperturbed residual $\mathbf{F}\left(\mathbf{h}^k\right)$ for SWR1 nonlinear iteration $k = 1$ based on the initial residual in equation 3–V1 is

$$\begin{bmatrix} 3.34683958405672 & 0.155127277391763 & - \\ 3.34683958405672 & 0.155127277391763 & 0.706811970303500 \\ - & 0.155127277391763 & 0.706811970303500 \end{bmatrix}. \tag{3–M3}$$

The stage perturbation, $\boldsymbol{\sigma}$, calculated using the initial stages in table 3–1 and equation 3–6 is

$$\begin{bmatrix} 2.250075340270996 \times 10^{8} \\ 2.086162567138672 \times 10^{8} \\ 1.892447471618652 \times 10^{8} \end{bmatrix}. \tag{3–V2}$$

The perturbed stage for SWR1 nonlinear iteration $k = 1$ is calculated by adding vector 3–V2 to the initial stages in table 3–1. The perturbed residual is calculated using equation 3–1 and the same function calls used to calculate matrix 3–M3. The perturbed residual, $\mathbf{F}\left(\mathbf{h}^k + \boldsymbol{\sigma}\right)$, for SWR1 nonlinear iteration $k = 1$ is

$$\begin{bmatrix} 3.34683927319263 & 0.155126966527673 & \\ 3.34683979901327 & 0.155127757130325 & 0.706811705521488 \\ & 0.155127108776336 & 0.706812311198105 \end{bmatrix}. \tag{3–M4}$$

The difference between the $\mathbf{F}\left(\mathbf{h}^k + \boldsymbol{\sigma}\right)$ and $\mathbf{F}\left(\mathbf{h}^k\right)$ is

$$\begin{bmatrix} 3.10864090025120 \times 10^{7} & 3.10864089997365 \times 10^{7} & \\ 2.14956549893230 \times 10^{7} & 4.79738562014775 \times 10^{7} & 2.64782011982767 \times 10^{7} \\ & 1.68615426981988 \times 10^{7} & 3.40894604988762 \times 10^{7} \end{bmatrix}. \tag{3–M5}$$

The approximate Jacobian, $\tilde{\mathbf{J}}$, for SWR1 nonlinear iteration $k = 1$ is calculated by dividing matrix 3–M5 by vector 3–V2. The approximate Jacobian, $\tilde{\mathbf{J}}$, is

$$\begin{bmatrix} 13.8157191852860 & 13.8157191852860 & \\ 10.3039213163512 & 22.9962213550295 & 12.6923000386783 \\ & 8.90991321698887 & 18.0134249719109 \end{bmatrix}. \quad (3\text{–M6})$$

Solving for the upgrade vector $\mathbf{v}^k$ using the approximate Jacobian in matrix 3–M6 and LU decomposition results in

$$(\mathbf{v} \mid \mathbf{h} \mid \mathbf{r}) \begin{bmatrix} 0.529567133726723 & 2.03956713372672 & 1.54055492735708 \\ 0.287318464716342 & 1.68731846471634 & 1.97443928987631 \\ 0.102877194028749 & 1.37287719402875 & 1.11749559483840 \end{bmatrix} \quad (3\text{–M7})$$

at the end of the SWR1 nonlinear iteration $k = 1$. Because the residual exceeds the specified TOLR criteria of 1×10^{-9} m³/s, additional SWR1 nonlinear iterations are required. At the end of SWR1 nonlinear iteration $k = 49$ the stage and residual were

$$(\mathbf{h} \mid \mathbf{r}) \begin{bmatrix} 2.01156378822003 & 4.408056648586012 \times 10^{7} \\ 1.80744207310212 & 1.723018794308473 \times 10^{8} \\ 1.54046654140130 & 4.820671914984587 \times 10^{7} \end{bmatrix}. \quad (3\text{–M8})$$

The residuals at the end of SWR1 nonlinear iteration $k = 49$ exceeded TOLR and were used as initial values for SWR1 nonlinear iteration $k = 50$. The cell-interface components used to calculate Q_M for each reach in equation 3–1 and resultant Q_M values at the beginning of the SWR1 nonlinear iteration $k = 50$ are

$$\left(\mathbf{A} \mid \mathbf{D}_M \mid \left(\frac{\Delta \mathbf{h}}{\Delta \mathbf{x}}\right) \mid \mathbf{Q}_M\right) \begin{bmatrix} 33.19006 & 885.56260 & 0.000204122 & 6.00000044080566 \\ 30.47909 & 737.18781 & 0.000266976 & 6.00000045803585 \\ 24.40467 & 453.93497 & 0.000540467 & 5.99999997596866 \end{bmatrix}. \quad (3\text{–M9})$$

The stage perturbation, σ, calculated using the stages in equation 3–M8 and equation 3–6 for SWR1 nonlinear iteration $k = 50$ is

$$\begin{bmatrix} 2.997463625997344 \times 10^{8} \\ 2.693298567983689 \times 10^{8} \\ 2.295474024714983 \times 10^{8} \end{bmatrix}. \quad (3\text{–V3})$$

Following the same process used to calculate matrix 3–M5 with the stage perturbation shown in vector 3–V3 results in the approximate Jacobian, $\tilde{\mathbf{J}}$,

$$\begin{bmatrix} 17.5385307875059 & 17.5385307875059 & \\ 11.8556984887318 & 26.2000709679253 & 14.3443724791934 \\ & 8.12959916661182 & 17.5996869249522 \end{bmatrix}. \quad (3\text{–M10})$$

Solving for $\mathbf{v}^k$ using the approximate Jacobian, $\tilde{\mathbf{J}}$, in matrix 3–M10 and LU decomposition results in

$$
(\mathbf{v} \mid \mathbf{h} \mid \mathbf{r}) \begin{bmatrix} 1.506731366142491 \times 10^{8} & 2.01156377315271 & 5.720476270454355 \times 10^{8} \\ 1.006624343063337 \times 10^{8} & 1.80744208316836 & 2.847491060720131 \times 10^{7} \\ 3.204044015712399 \times 10^{8} & 1.54046657344174 & 6.255942963662164 \times 10^{8} \end{bmatrix}
$$

(3–M11)

at the end of the SWR1 nonlinear iteration $k = 50$. The residual at the end of SWR1 nonlinear iteration $k = 50$ exceeds the specified TOLR criteria. Because this example is a steady-state simulation, the initial heads at the beginning of the time step (table 3.1) are reset to the heads at the end of SWR1 nonlinear iteration $k = 50$, and addition MODFLOW Picard iterations are performed until TOLR is satisfied and the change in volume between subsequent SWR1 outer iterations is less than TOLR. For transient problems, the SWR1 solution would be considered non-converged at the end of SWR1 nonlinear iteration $k = 50$, and the simulation would be terminated unless the non-convergence continuation flag (USE_NONCONVERGENCE_CONTINUE option specified) was set in the SWR1 input dataset.

The final stage and residual were

$$
(\mathbf{h} \mid \mathbf{r}) \begin{bmatrix} 2.01156374247530 & 8.488321157074097 \times 10^{12} \\ 1.80744204261062 & 9.042544490966975 \times 10^{12} \\ 1.54046654393978 & 9.277023593767808 \times 10^{12} \end{bmatrix}
$$

(3–M12)

Convergence for the TOLR criteria at all reaches required 2 MODFLOW Picard iterations and 69 calls to the LU decomposition solver in SWR1. A total of 15 MODFLOW Picard iterations, or 13 additional MODFLOW Picard iterations, were required to reduce volume changes to less 1×10^{-9} m³/s.

Appendix 4.   SWR1 Process Preprocessor for Reach Segmentation and Connectivity

SWR1 Preprocessor Introduction

One of the more complicated tasks involved in creating an SWR1 Process is the preparation of SWR1 data item 4a and 4b (app. 2), which contains information about the cell, length, and connections of every reach in the model. The SWR1 Process preprocessor is designed to simplify this task by generating this dataset in a simple and intuitive way, using input data in formats that a user of MODFLOW would be likely to have.

Preprocessor Inputs

The preprocessor requires two input files and five pieces of supplemental information. A MODFLOW discretization file provides the number and spacing of rows and columns, and an Esri shapefile provides information about waterway geometries and, through shapefile attributes, waterway identifiers and inter-waterway connectivity. The shapefile should be in a projected (Cartesian) coordinate system (for example, State Plane, Albers equal area) that is appropriate for the geographic area covered by the model domain. A State Plane coordinate system would be appropriate for model domains contained in one State Plane zone. Albers equal area or Lambert conformal conic projections should be used for models that cover multiple State Plane zones or multiple states. Additionally, the program requires an origin location for referencing the discretization grid to the shapefile coordinate system, a rotation angle to allow the grid to be referenced to this coordinate system non-orthogonally, a minimum reach group length, a specification of the treatment of the reach group length, and a preferred ordering direction for numbering the SWR1 reaches. **The MODFLOW discretization file must be a complete and valid file** (Harbaugh, 2005, p. 8-11 – 8-13).

To provide reach identifiers and reach connectivity, the shapefile must contain three attributes: NAME, NCONN, and ICONN. These attributes must be either character strings or integer values, and the attribute names and values (if provided as character strings) are not case-sensitive. All of the attributes must be of the same type. For example, if name is an integer attribute, then NCONN and ICONN must also be integer attributes.

NAME must contain a value to identify the feature, and multiple features with the same NAME value will be treated as a single feature. If this attribute is an integer attribute, all values must be positive. If this attribute is provided as a character string attribute, all values must be non-empty.

NCONN must contain a value to identify the number of connections for the feature, and ICONN must be a comma-separated or whitespace-separated list of connections for the feature. NCONN can be set to zero if the feature is not connected to another feature; in this case, ICONN is not used for the feature. In cases where a feature has a non-zero NCONN value, the only valid values for the ICONN attribute are values that are specified in the NAME field of another feature in the shapefile.

The point location and rotation angle must be numerical values (decimal values are allowed). The point location must be in the same coordinate system as the shapefile, and it specifies the southwest corner of the southwest cell of the grid. The rotation angle must be in degrees, and it specifies counterclockwise rotation of the grid about the point location (fig. 4–1).

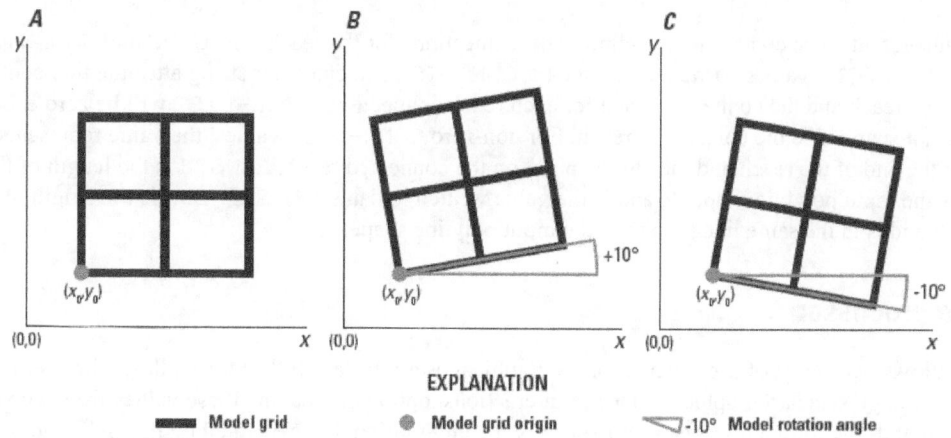

Figure 4–1. Examples of three 2-by-2 discretization grids with rotation angles of *A*, 0; *B*, 10; and *C*, -10 degrees, respectively.

The reach-group length must be a numerical value (a decimal value is allowed) and it must use the same units as the shape-file coordinate system. This argument has a different meaning depending on the chosen reach-group length option.

If the "intersection" reach-group length option is chosen, the reach-group length argument is not used. Instead, each reach will be defined to be a unique reach group determined by intersection of the input polyline shapefile with the MODFLOW grid.

If the "exact" reach-group length option is chosen, each reach group is made to be exactly this length, with the exception of, possibly, the last (the reach group associated with the last vertex of the polyline) reach group. The last reach group will be between 0.5 (inclusive) and 1.5 (exclusive) times the specified reach group length.

If the "equal" reach-group length option is chosen, the reach-group length value is adjusted for each input polyline so that the actual length of every reach group derived from that polyline is the closest length to the specified length that divides the total length of the polyline by a whole number.

If the preferred ordering direction option is used, reaches will be numbered starting in the specified quadrant. The preferred ordering option will proceed along a polyline to a connection and continue numbering along connected polylines starting with the polyline that is closest to the preferred ordering direction in the case of multiple connected polylines. Available preferred ordering direction options include "top left", "lower left", "top right", "lower right", and "none". Specification of the "none" preferred ordering direction option results in assignment of reach numbers based on the order of polyline entries (FID attribute) and polyline vertices in the input shapefile. The preferred ordering direction option can result in a more logical ordering of reach numbers that is independent of the order of polyline entries in the input shapefile.

Preprocessor Outputs

The preprocessor produces three output files in the current working directory. An Esri shapefile of the discretization grid is written to grid.shp. An Esri shapefile of the input polylines, after discretization, is written to polylines.shp. Both shapefiles are assigned the same coordinate system and georeferencing information as the input shapefile. A text file containing the SWR1 Process input item 4a and 4b is written to dataset.txt.

The grid.shp file contains polygons of the discretized grid described by the discretization file, point location, and rotation angle. The shapefile contains two attributes, ROW and COLUMN. These contain the row and column number of each cell. The row numbers start at index 1 at the west side of the grid, and they increase to the east. The column numbers start at index 1 at the north side of the grid, and they increase to the south.

The polylines.shp file contains the polylines, discretized to the specified grid. The shapefile contains 11 attributes, SRC_PLY, SRC_REACH, COLUMN, ROW, REACH, RCH_GRP, CONN, NCONN, CONDIST, LENGTH, and GRP_LEN. SRC_PLY is an integer value that indicates the polyline entry (FID attribute) in the input shapefile that was intersected with the model grid to define the discretized SWR1 reach. SRC_REACH is the name of the polyline entry in the input shapefile. COLUMN and ROW are integer attributes containing the row and column of the model grid cell that contains the reach. REACH is a unique integer value for the reach. RCH_GRP is an integer value of the reach group containing the reach. The values defined for the REACH and RCH_GRP attributes are the same as the reach number and reach group that is written to SWR1 input item 4a. CONN is a character string attribute that contains the values from REACH and ICONN in the following format:

REACH(ICONN1,ICONN2,…,ICONN(NCONN)). ICONN values are provided to allow for inspection of reach connections in GIS software such as ArcMap.

NCONN is an integer attribute containing the number of connections for the reach. The CONN and NCONN attribures are the same as the ICONN and NCONN values written to dataset 4. CONN_DIST is a character string attribute that contains the distance between the end of the reach and the connected reach for each reach connection. CONN-DIST will be zero except when the end of a polyline is not snapped to the connecting reach. For non-zero CONN-DIST values, the value represents the physical separation between the end of the reach and the closest point on the connected reach. LENGTH is the length of the reach in the same linear units as the input polyline shapefile and is the value written to dataset 4. GRP_LEN is the length of the reach group containing the reach and is in the same linear units as the input polyline shapefile.

Running the Preprocessor

The program allows two forms of user interaction. A graphical user interface (GUI) form allows the user to specify the program arguments through standard graphical interface interactions, optionally loading these values from a configuration file. A non-interactive command-line form allows the user to specify a configuration file, generated by the GUI, and immediately initiate the program's processing function.

Interactive Graphical Interface

The interactive graphical interface provides standard GUI elements for the specification of arguments. It also provides a menu option for loading arguments from and saving arguments to a configuration file. The graphical interface may be invoked by double-clicking on the application icon:

SWRPre

The graphical interface has a standard Windows look and functionality and is shown in figure 4–2.

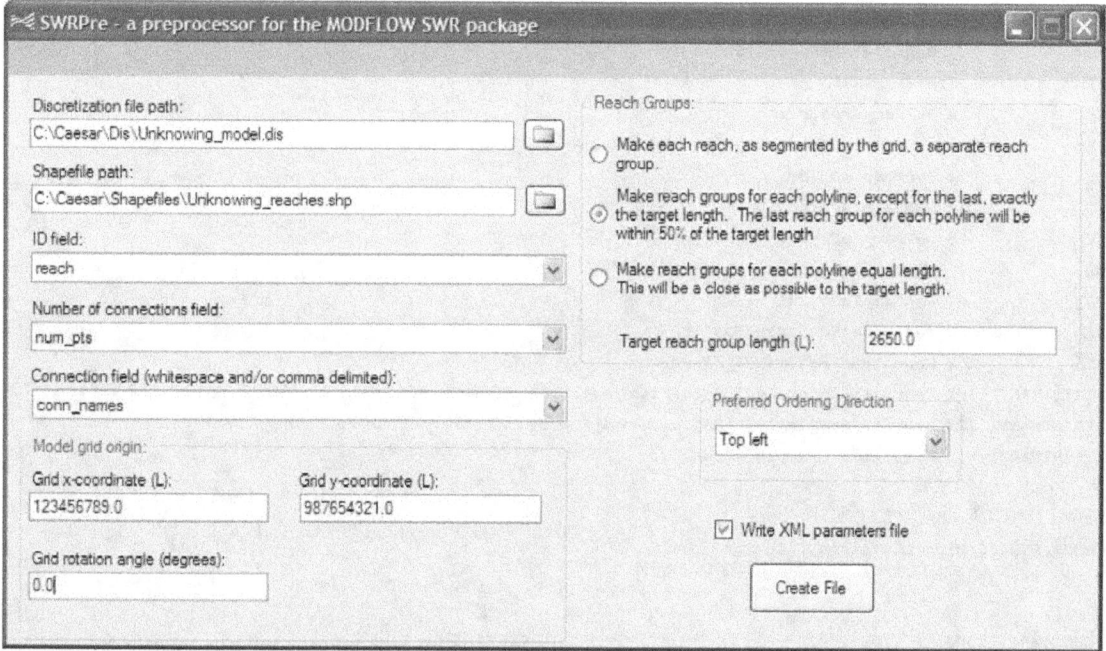

Figure 4–2.   Example of the graphical user interface for the SWR1 Preprocessor.

Non-interactive command-line interface

The non-interactive command-line interface requires specification of all arguments from the command line. It may be invoked as follows:

```
SWRPre XML-config-file-path
```

Appendix 5.   List of Selected Input Datasets for Test Simulation 5

The NAM, DIS, BAS, LPF, UZF1, SWR1, and the external time series input datasets for test simulation 5 are presented herein to provide users with a quick reference for setting up a MODFLOW model that uses the SWR1 Process. Gaps in the input datasets are indicated by an ellipsis. The test simulation 5 SWR1 dataset uses an external file (SSWR_RDTABDATA subroutine) to specify stage values for constant-stage reaches 1-6, two-dimensional rainfall data (RAIN) from external files (U2DREL subroutine), list data (SSWRLSTRD subroutine) to specify evaporation data (EVAP), and constant lateral flow data (QLATFLOW — U2DREL subroutine). Test simulation 5 also includes a variety of reach geometry types (IGEOTYPE) and surface-water control structures types (ISTRTYPE) and uses a number of reach conductance options (IGCNDOP). A complete set of the files for test simulation 5 are available for distribution over the Internet as discussed in the Preface. Some brief annotations have been added as comments within the SWR1 dataset to help the reader understand the purpose of various sections of the input dataset. Comments in the SWR1 dataset are identified with a "#" in column 1. Font sizes of the input datasets have been reduced so that lines will fit within page margins.

File name: SWRTestSimulation05.nam

```
LIST            7 Results\SWRSample05.lst
BAS6            1 SWRSample05.bas
DIS            29 SWRSample05.dis
LPF            11 SWRSample05.lpf
OC             22 SWRSample05.oc
PCG            19 SWRSample05.pcg
UZF            18 SWRSample05.UZF
GHB            16 SWRSample05.ghb
SWR            61 SWRSample05.swr
DATA          201 ref\ConstantStage.dat
DATA          102 Results\SWRSample05_Stage.csv
DATA(BINARY)  101 Results\SWRSample05_GroupFlow.flow
DATA(BINARY)  103 Results\SWRSample05_QAQ.bin
DATA(BINARY)  111 Results\SWRSample05.hds
```

File name: SWRTestSimulation05.dis

```
# MODFLOW2005 Discretization File
 2   6   6  367  4   2
 0  0
    CONSTANT    304.80 (10E12.4)                 0      DELTA-X or C
    CONSTANT    304.80 (10E12.4)                 0      DELTA-Y or R
    INTERNAL      1.00 (10e12.4)                -1      TOP of Model
         1.530        1.724        1.824        1.524        1.424        1.224
         1.530        1.524        1.624        1.324        1.224        1.224
         1.530        1.624        1.424        1.224        0.3048       1.124
         1.530        1.524        1.424        1.324        0.3048       1.024
         1.530        1.624        1.524        1.424        0.924        0.824
         1.530        1.724        1.624        1.624        1.124        1.424
    CONSTANT      0.25 (10e12.4)                -1      BOTTOM of Layer    1
    CONSTANT     -20.0 (10e12.4)                -1      BOTTOM of Layer    2
 1.00      6   1.00       TR   STRESS PERIOD 001
 1.00      6   1.00       TR   STRESS PERIOD 002

 . . .
```

DIS Input data for stress periods 3-366 are not shown.

```
 1.00      6   1.00       TR   STRESS PERIOD 367
```

File name: SWRTestSimulation05.bas

```
# MODFLOW2005 Basic Package
#

   CONSTANT        1 (FREE)                    -1    IBOUND Layer    1
   CONSTANT        1 (FREE)                    -1    IBOUND Layer    2
9.9900e+02
   INTERNAL      1.00 (FREE)                    0    STARTING HEADS Layer    1
       1.534        1.362      1.413     1.158      1.043      0.8706
       1.534        1.184      1.188     0.9792     0.8792     0.8554
       1.534        1.257      1.153     1.006      0.9445     0.7678
       1.534        1.180      1.144     0.9598     0.9259     0.6595
       1.534        1.269      1.151     1.026      0.5374     0.5005
       1.534        1.357      1.243     1.186      0.7546     0.9789
   INTERNAL      1.00 (FREE)                    0    STARTING HEADS Layer    2
       1.530        1.363      1.402     1.157      1.039      0.8739
       1.527        1.196      1.187     0.9858     0.8854     0.8547
       1.528        1.258      1.153     1.006      0.9405     0.7711
       1.526        1.190      1.141     0.9648     0.9142     0.6645
       1.528        1.272      1.153     1.021      0.5595     0.5311
       1.530        1.356      1.242     1.175      0.7637     0.9646
```

File name: SWRTestSimulation05.lpf

```
# MODFLOW2000 Layer Property Flow (LPF) Package
        0  -1.000000e+030        0
 1 1
 0 0
 1.000000e+000 1.000000e+000
 1 1
 1 1
 0.1 4 1
   INTERNAL     10.00 (FREE)                    0    HYDRAULIC CONDUCTIVITY Layer    1
 1.0  1.0  1.0  1.0  1.0  1.0
 1.0  0.5  0.5  1.0  1.0  1.0
 1.0  1.0  0.5  1.0  1.0  1.0
 1.0  0.5  0.5  1.0  1.0  1.0
 1.0  1.0  1.0  1.0  1.0  1.0
 1.0  1.0  1.0  1.0  1.0  1.0
        0    10.000(10e12.4)                    0    VERTICAL HYDRAULIC CONDUCTIVITY Layer    1
        0    1.0e-05(10e12.4)                   0    PRIMARY STORAGE Layer    1
        0     0.200(10e12.4)                    0    Sy Layer    1
        0     0.100(10e12.4)                    0    WETDRY Layer 1
        0     5.000(10e12.4)                    0    HYDRAULIC CONDUCTIVITY Layer    2
        0    10.000(10e12.4)                    0    VERTICAL HYDRAULIC CONDUCTIVITY Layer    2
        0    1.0e-05(10e12.4)                   0    PRIMARY STORAGE Layer    2
        0     0.200(10e12.4)                    0    Sy Layer    2
        0     0.100(10e12.4)                    0    WETDRY Layer 2
```

File name: SWRTestSimulation05.uzf

```
# MODFLOW2005 UZF Input File
#NUZTOP IUZFOPT IRUNFLG IETFLG IUZCB1 IUZFCB2 [NTRAIL2 NSETS2] NUZGAG SURFDEP
       1       2       1      1      0       0      20      20      0    0.50
INTERNAL      1        (FREE)      -1              #IUZFBND
 0 1 1 1 1 1
 0 1 1 1 1 1
```

```
0 1 1 1 0 1
0 1 1 1 0 1
0 1 1 1 1 1
0 1 1 1 1 1
INTERNAL    1        (FREE)    -1                    #IRUNBND
 000000 100007 100008 100014 100015 100015
 000000 100007 100008 100014 100015 100015
 000000 100011 100011 100014 000000 100016
 000000 100009 100010 100014 000000 100018
 000000 100009 100010 100017 100017 100018
 000000 100009 100010 100019 100019 100018
CONSTANT    3.500             1.000 (FREE)    -1     #EPS
CONSTANT    0.25              1.000 (FREE)     1     #THTS
CONSTANT    0.15              1.000 (FREE)     1     #THTI
1                                                   #NUZF1
OPEN/CLOSE ref\Rainfall_001.ref 1.00 (FREE)    -1   #FINF  STRESS PERIOD 001
1                                                   #NUZF2
CONSTANT  2.6249E-03  1.000 (FREE)             1    #PET
1                                                   #NUZF3
CONSTANT 0.25 1.000   (FREE)                   1    #ET Ext Depth
1                                                   #NUZF4
CONSTANT 0.05 1.000 (FREE)                     1    #Ext Water Content
1                                                   #NUZF1 STRESS PERIOD 002
OPEN/CLOSE ref\Rainfall_002.ref 1.0 (FREE)     -1   #FINF  STRESS PERIOD 002
-1                                                  #NUZF2 STRESS PERIOD 002
-1                                                  #NUZF3 STRESS PERIOD 002
-1                                                  #NUZF4 STRESS PERIOD 002


. . .
```

UZF1 Input data for stress periods 3-366 are not shown.

```
1                                                   #NUZF1 STRESS PERIOD 367
OPEN/CLOSE ref\Rainfall_367.ref 1.0 (FREE)    -1    #FINF  STRESS PERIOD 367
-1                                                  #NUZF2 STRESS PERIOD 367
-1                                                  #NUZF3 STRESS PERIOD 367
-1                                                  #NUZF4 STRESS PERIOD 367
```

File name: `SWRTestSimulation05.swr`

```
# SURFACE WATER ROUTING (SWR1) PROCESS TEST DATASET
# TITLE - SWR TEST SIMULATION 5
# DATASET 1 - DIMENSIONS AND PRELIMINARIES
# NREACHES ISWRONLY ISWRCBC ISWRPRGF ISWRPSTG ISWRPQAQ ISWRPQM ISWRPSTR ISWRFRN Option
        19        0       0     -101      102     -103        0        0       1     SWROPTIONS


# DATASET 1B - SWR1 OPTIONS
PRINT_SWR_TO_SCREEN
USE_TABFILES
USE_NONCONVERGENCE_CONTINUE
END


# DATASET 2 - SOLUTION CONTROLS
#DLENCONV TIMECONV    RTINI      RTMIN      RTMAX  RTPRN RTMULT NTMULT DMINGRAD DMNDEPTH DMAXRAI DMAXSTG DMAXINF
     1.0   86400. 6.9444E-03 3.4722E-04 6.9444E-03 1.0E+00  1.01    10 1.0E-12  1.0E-03 0.15240    0.50     0.0
```

```
# DATASET 3 - SOLVER PARAMETERS
#ISOLVER NOUTER NINNER IBT    TOLS    TOLR    TOLA DAMPSS DAMPTR IPRSWR MUTSWR IPC NLEVELS DROPTOL IBTPRT
       2    031    100 10 1.0E-09   86.40   0.010    1.0    1.0      0      0   3                    0
```

```
# DATASET 4A - REACH DATA
#                                  LAY ROW  COL
#IREACH IROUTETYPE IRGNUM KRCH IRCH JRCH   RLEN
      1          1     01    1    1    1  304.8
      2          1     01    1    2    1  304.8
      3          1     01    1    3    1  304.8
      4          1     02    1    4    1  304.8
      5          1     02    1    5    1  304.8
      6          1     02    1    6    1  304.8
      7          2     03    1    2    2  350.0
      8          2     03    1    2    3  350.0
      9          2     03    1    4    2  340.0
     10          2     03    1    4    3  460.0
     11          2     03    1    3    3  100.0
     12          2     03    1    3    3  250.0
     13          2     03    1    3    3  150.0
     14          2     03    1    3    4  380.0
     15          1     04    1    3    5  304.8
     16          1     04    1    4    5  304.8
     17          3     05   -1    5    5  790.0
     18          3     06   -1    5    6  325.0
     19          1     07   -1    6    5  150.4
```

```
# DATASET 4B - REACH CONNECTIVITY
#IREACH NCONN ICONN
      1     1     2
      2     3     1 3 7
      3     1     2
      4     2     5 9
      5     2     4 6
      6     1     5
      7     2     2 8
      8     2     7 11
      9     2     4 10
     10     2     9 12
     11     3     8 12 13
     12     3     10 11 13
     13     3     11 12 14
     14     2     13 15
     15     2     14 16
     17     3     16 18 19
     18     1     17
     19     1     17
```

```
# DATASET 4C - TABULAR DIMENSION DATA
#NTABS
      1
```

```
# DATASET 4D - TABULAR SPECIFICATION DATA
#CTABTYPE ITABUNIT   CINTP NTABRCH ITABRCH(1:NTABRCH)
   STAGE      201    NONE      6 1 2 3 4 5 6


# DATASET 5 - STRESS PERIOD 1
# ITMP IRDBND IRDRAI IRDEVP IRDLIN IRDGEO IRDSTR IRDSTG IPTFLG
    1     19    -19     19    -19     19      6     19      1


# DATASET 6 - BOUNDARY DATA
# IBNDRCH ISWRBND
        01      -1
        02      -1
        03      -1
        04      -1
        05      -1
        06      -1
        07       1
        08       1
        09       1
        10       1
        11       1
        12       1
        13       1
        14       1
        15       1
        16       1
        17       1
        18       1
        19       1


# DATASET 7B - RAINFALL DATA
OPEN/CLOSE ref\Rainfall_001.ref 1.00 (FREE)    -1        #STRESS PERIOD 001


# DATASET 8A - EVAPORATION DATA
        01      3.7498E-03
        02      3.7498E-03
        03      3.7498E-03
        04      3.7498E-03
        05      3.7498E-03
        06      3.7498E-03
        07      3.7498E-03
        08      3.7498E-03
        09      3.7498E-03
        10      3.7498E-03
        11      3.7498E-03
        12      3.7498E-03
        13      3.7498E-03
        14      3.7498E-03
        15      3.7498E-03
        16      3.7498E-03
        17      3.7498E-03
        18      3.7498E-03
        19      3.7498E-03
```

```
# DATASET 9A - LATERAL FLOW DATA
  CONSTANT 0.0000E-00 (11e12.4)                    +1


# DATASET 10 - GEOMETRY ASSIGNMENT DATA
# DATASET 8A
INTERNAL
#IGMODRCH IGEONUMR GZSHIFT
       01        1     0.0
       02        1     0.0
       03        1     0.0
       04        1     0.0
       05        1     0.0
       06        1     0.0
       07        2     0.0
       08        2     0.0
       09        3     0.0
       10        3     0.0
       11        2     0.0
       12        2     0.0
       13        2     0.0
       14        2     0.0
       15        4     0.0
       16        4     0.0
       17        6     0.0
       18        7     0.0
       19        5     0.0


# DATASET 11A - GEOMETRY DATA
# IGEONUM IGEOTYPE IGCNDOP GMANNING NGEOPTS GWIDTH GBELEV GSSLOPE    GCND        GLK GCNDLN GETEXTD
        1        5       0     0.25                                9.2E+04                      0.25
        2        2       1     0.03           30.00 0.3048    1.0  2.50E-01
        3        1       1     0.03           30.00 0.3048         2.50E-01
        4        5       1     0.25                                2.00E-02             0.25
        7        3       1     0.03       8                        5.00E-01
# DATASET 11B - FOR IGEONUM 7
'OPEN/CLOSE' 'IrregularCrossSection_Reach18.dat'
# DATASET 11A - IGEONUM 5
        5        4       1     0.05       7                        1.00E-02
# DATASET 11B - FOR IGEONUM 6
'OPEN/CLOSE' 'SVAPCrossSection_Reach19.dat'
# DATASET 11A - IGEONUM 6
        6        3       3     0.10       8                        1.00E-01  63.50
# DATASET 11B - FOR IGEONUM 7
'OPEN/CLOSE' 'IrregularCrossSection_Reach17.dat'


# DATASET 12 - STRUCTURE ASSIGNMENT DATA
#ISMODRCH NSTRUCT
       02        1
       04        1
       14        2
       16        1
       18        1
       19        1
```

```
# DATASET 13A
#ISTRRCH ISTRNUM ISTRCONN ISTRTYPE NSTRPTS STRCD STRCD2 STRCD3 STRINV STRINV2 STRWID STRWID2 STRLEN STRMAN STRVAL ISTRDIR
    02      1        7        6             0 61          0 5   1 535        45 00                        0 0      0
    04      1        9        6             0 61          0 5   1 535        45 00                        0 0      0
    14      1       15        9             0 61    0 61  0 5   1 000        30 00                        0 0      0
# DATASET 13B for structure 1 in reach 14
#CSTROVAL ISTRORCH ISTROQCON ISTRLO STRCRIT STRCRITC  STRRT STRMAX    CSTRVAL
   STAGE      14                  GE  1 250       0 0  108 0    0 25
# DATASET 13A
#ISTRRCH ISTRNUM ISTRCONN ISTRTYPE NSTRPTS STRCD STRCD2 STRCD3 STRINV STRINV2 STRWID STRWID2 STRLEN STRMAN STRVAL ISTRDIR
    14      2       15        3                                                                            0 0
# DATASET 13B for structure 2 in reach 14
#CSTROVAL ISTRORCH ISTROQCON ISTRLO STRCRIT STRCRITC  STRRT STRMAX    CSTRVAL
   STAGE      14                  GE  1 750       0 0  450 0 21600
# DATASET 13A
#ISTRRCH ISTRNUM ISTRCONN ISTRTYPE NSTRPTS STRCD STRCD2 STRCD3 STRINV STRINV2 STRWID STRWID2 STRLEN STRMAN STRVAL ISTRDIR
    16      1       17        6             0 61          0 5   1 000        40 00                        0 0      0
    18      1        0        6             0 61          0 5   0 500        45 00                        0 0      0
    19      1       17        8             0 61          0 5   0 450        10 00                        0 25     0
# DATASET 13B for structure 1 in reach 19
#CSTROVAL ISTRORCH ISTROQCON CSTRLO STRCRIT STRCRITC  STRRT STRMAX    CSTRVAL
   STAGE      19                  GE  0 750       0 0  108 0    0 25


# DATASET 14A - REACH STAGE DATA
INTERNAL
#IRCHSTG  STAGE
      01  1.537
      02  1.537
      03  1.537
      04  1.537
      05  1.537
      06  1.537
      07  1.219
      08  1.219
      09  1.219
      10  1.219
      11  1.219
      12  1.219
      13  1.219
      14  1.219
      15  1.000
      16  1.000
      17  0.609
      18  0.609
      19  1.000


#DATASET 5 - STRESS PERIOD 002
# ITMP IRDBND IRDRAI IRDEVP IRDLIN IRDGEO IRDSTR IRDSTG IPTFLG
     1     0     -1      0      0      0      0      0      0
# DATASET 7B - RAINFALL DATA
OPEN/CLOSE ref\Rainfall_002.ref 1.0 (FREE)   -1          #STRESS PERIOD 002


  . . .
```

SWR1 Input data for stress periods 3-366 are not shown.

```
#DATASET 5 - STRESS PERIOD 367
# ITMP IRDBND IRDRAI IRDEVP IRDLIN IRDGEO IRDSTR IRDSTG IPTFLG
    1     0    -1     0      0      0      0      0      0
# DATASET 7B - RAINFALL DATA
OPEN/CLOSE ref\Rainfall_367.ref 1.0 (FREE)    -1          #STRESS PERIOD 367
```

File name: `ConstantStage.dat`

```
#--EXTERNAL TIMESERIES FILE USED TO
#  DEFINE THE STAGE VALUES FOR
#  CONSTANT STAGE REACHES 1-6
#  OPTIONAL ITEM 1
OFFSET 0.0 SCALE 1.0
#  ITEM 2
#  SIMTIME    STAGE
        0    1.537
        1    1.537
        2    1.536
        3    1.535
        4    1.535
        5    1.535
        6    1.535
        7    1.536
        8    1.535
        9    1.535
       10    1.535
```

. . .

Data for simulation days 11-359 are not shown.

```
      360    1.535
      361    1.535
      362    1.535
      363    1.535
      364    1.535
      365    1.540
      366    1.540
      367    1.537
      368    1.537
```